P9-DNW-681

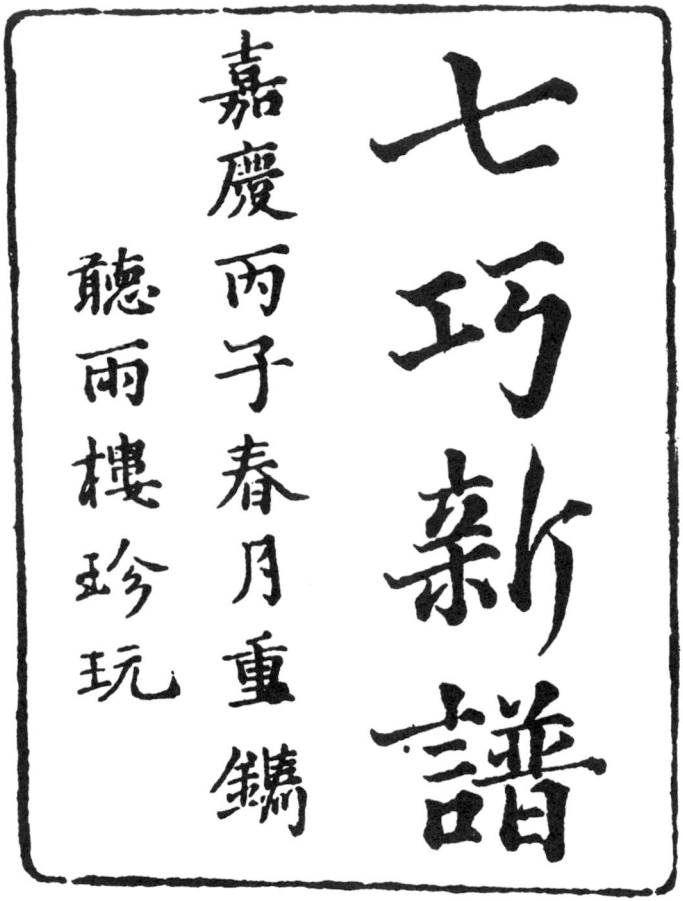

The title page of a Chinese tangram book (see page 59).

TANGRAMS—

330 Puzzles

by Ronald C. Read

Dover Publications, Inc., New York

Copyright © 1965 by Dover Publications, Inc.

All rights reserved.

Tangrams—330 Puzzles is a new work, first published by Dover Publications, Inc., in 1965.

Standard Book Number: 486-21483-4

Library of Congress Catalog Card Number 65-26651

Manufactured in the United States by Courier Corporation
21483432
www.doverpublications.com

Table of Contents

		page
	Introduction	1
1.	Letters and Numbers	9
2.	Animals	15
3.	Mankind	25
4.	Around the House	35
5.	Boats and Bridges	39
6.	Stories and Pictures	43
7.	A Little Mathematics	53
8.	Chinoiserie	59
9.	Paradoxes and Illusions	69
10.	Double Tangrams	77
11.	The Fifteen-Piece Puzzle	81
	Solutions	93

Introduction

THE seven-piece puzzle, or tangram as it is usually known in the West, has come to us from China. It is in all probability the original "Chinese puzzle"—the proverbial prototype of all that is perplexing and tricky. This puzzle resembles the familiar jigsaw puzzle in that it is concerned with the fitting together of geometrical shapes, yet in all other respects the two puzzles are completely different; for whereas the jigsaw puzzle consists of a large number of pieces of rather complicated shapes, which have to be fitted together in a unique way, the tangram consists of only seven pieces of simple shapes, and the whole charm of the puzzle lies in the extraordinary variety of ways in which these pieces can be put together.

The seven pieces which make up the tangram can be cut from a single square, as shown in Figure 1. There are thus two small triangles, one medium-sized triangle, and two large triangles, in addition to a square and a lozenge-shaped piece.

FIGURE 1

The medium-sized triangle and the square and the rhomboid are all twice the area of one of the small triangles; each of the large triangles is four times the area of a small triangle. All the angles in these pieces are either right angles or angles of 45° or 135°. A set of tangram pieces can easily be made from cardboard. However, the enthusiast will probably want to make his set from wood, plastic or some other more durable material.

With a set of seven tangram pieces, the reader can make (if he tries hard enough) all the outlines that are to be found in the following pages except those of the last two chapters, and a quick glance through these pages will reveal the surprisingly large number of different outlines that can be constructed from such a small number of pieces. This is what has made the tangram popular in the country of its origin and also in the West. From time to time it has been revived on both sides of the Atlantic and has enjoyed a certain vogue, even though it has never achieved the status of such established pastimes as checkers or tic-tac-toe.

The tangram has a reputation for being very old, and this may well be, though one should beware the common tendency to ascribe exaggerated age to all things Chinese. I have in my possession two books of tangrams printed in China in 1813 and 1823. In the preface to one of these the author says that "the origin of the seven-piece puzzle is not known . . ." This would suggest that the puzzle was even then regarded as old, though just how old is a matter for conjecture. Despite this, I have been quite unable to find any Chinese books on tangrams, or even any reference to this kind of puzzle, dating earlier than the beginning of the nineteenth century, at which time a large number of books on the subject seem to have been published.

There was, apparently, a resurgence of interest in the tangram in China around this time, resulting in the publication in book form of what may previously have been handed down only by word of mouth. It should be mentioned here that the Chinese have always regarded the seven-piece puzzle as predominantly a game for children and women, and therefore perhaps hardly a worthy subject for books. I venture to hope that this is not true of the present book, which contains many subtleties devised by occidental puzzlers that are not to be found in the Chinese texts.

It was during the first half of the nineteenth century that the tangram became known in Western countries, and books about it started to appear. At first, these books were nothing more than exact copies of Chinese texts, with the addition of hard covers and some sort of English preface explaining—often not very clearly—what the puzzle was about. A copy of one such book, now in the British Museum, bears on its cover the following:

A Grand Eastern Puzzle

THE following Chineze Puzzle is recommended to the Nobility, Gentry and others, being superior to any hitherto invented for the Amusement of the Juvenile World, to whom it will afford unceasing recreation and information: being formed on Geometrical principles, it may not be considered as trifling to those of mature years, exciting interest, because difficult and instructive, imperceptibly leading the mind on to invention and perseverance. —— The Puzzle consists of five triangles, a square, and a rhomboid, which may be placed in upwards of THREE HUNDRED and THIRTY Characters, greatly resembling MEN, BEASTS, BIRDS, BOATS, BOTTLES, GLASSES, URNS, etc. The whole being the unwearied exertion of many years study and application of one of the Literati of China, and is now offered to the Public for their patronage and support.

ENTERED AT STATIONER'S HALL

Published and sold by
C. DAVENPORTE and CO.
No. 20, Grafton Street, East Euston Square.

Later on, publications appeared in Europe and America which were more original, but which still depended very heavily on the earlier sources. The same outlines appear, often in the same order, and often with mistakes (common in the Chinese books) faithfully copied! One such publication which I have in my possession was printed in Philadelphia in 1844. Its derivation from the earlier Chinese books is quite apparent, but it is greatly inferior to them. Approximately three hundred tangrams are given, but they are jumbled together any which way, and no indication is given of what the outlines are supposed to represent. This reduces the puzzle to the mere construction of meaningless geometrical patterns, which must have provided rather dull entertainment.

Lewis Carroll had in his library a book of tangrams entitled *The Fashionable Chinese Puzzle*, which seems to have been very similar to the American publication just mentioned. The great English puzzler H. E. Dudeney, who later acquired this book, describes it as follows:

> It contains three hundred and twenty-three Tangram designs, mostly nondescript geometrical figures, to be constructed from the seven pieces. . . . There is no date, but the following note fixes the time of publication pretty closely: "This ingenious contrivance has for some time past been the favourite amusement of the ex-Emperor Napoleon, who, being now in a debilitated state and living very retired, passes many hours a day in thus exercising his patience and ingenuity."

This reference to Napoleon's interest in the puzzle occurs also in the preface to the Philadelphia publication, but it would seem unlikely that there is any truth in the assertion. Napoleon's biographers, writing of his exile on St. Helena, mention that he was interested (though not skilled) in billiards, chess and *reversi*; nothing is said of anything at all resembling the tangram. Still, it is not impossible that Napoleon was acquainted with this form of entertainment, and the story of the Emperor whiling away his declining years in this manner, if not true, is at all events a pleasing enough fabrication.

All this goes to show that historical information about the tangram is almost nonexistent; but though fact is in short supply there is fiction in abundance. A detailed and highly colored "tangram legend" has come into existence, mainly due to the efforts of Sam Loyd, that great American puzzle expert of the turn of the century, aided and abetted by H. E. Dudeney in England. Sam Loyd's contribution to the literature of tangrams (and a very notable one it is, too) was a book called *The 8th Book of Tan*, which was published by Loyd himself in 1903. In this book Sam Loyd brought his own peculiar genius to bear on all aspects of the puzzle; in addition to creating hundreds of completely new shapes, he also dreamt up some paradoxical twists and quirks that had never been thought of before.

Along with these tangrams Sam Loyd supplied a running commentary, in which he gave what has every appearance of being a remarkably precise and carefully documented account of the history of the puzzle, its esoteric religious significance and symbolism, and its relation to the theorem of Pythagoras and other parts of Euclid's geometry, not to mention much additional scholarly and erudite information.

In point of fact, however, this painstaking work of scholarship is one delightful spoof from beginning to end! One example will suffice to show the sort of tall yarn that Sam Loyd pulls out of the hat. He quotes the researches of the "famous Professor Challenor" (a purely fictitious character, as far as I have been able to discover), who is said to have made a special study of the seven-piece puzzle. According to these researches the puzzle originated with a monumental work in seven volumes entitled *The Seven Books of Tan*, which was compiled in China some four thousand years ago. These books are described as "rare" (an understatement), and mention is made of one of them, printed in gold leaf upon parchment, that was found in Peking. Now, none of this is even remotely plausible. In 2000 B.C. the Chinese were still at the stage in which their literary achievements mainly consisted of crude pictographs scratched on tortoise shells; they had certainly not progressed even to writing on thin strips of bamboo—the way in which the Confucian classics, for example, were originally recorded (around 500 B.C.). A work of seven volumes would be quite out of the question at that time, and a book printed in gold leaf on parchment would be most unusual (to say the least) at any time.

I must beg the reader's indulgence if he considers that this is all too obvious to be worth the trouble of explanation. I take the trouble because it would seem that H. E. Dudeney accepted these tall stories at their face value; if he did not, he at least repeated them with such seriousness and seeming credence that his readers, faced with the unanimity of two famous authorities, might well believe them to be true.

This, of course, is exactly what Sam Loyd would like. His spoofing is clearly deliberate, and is completely in keeping with the good-natured leg-pulling that lends special appeal to so many of his puzzles.*

Dudeney's long preamble to Problem 169 (A Tangram Paradox) in his *Amusements in Mathematics*† repeats the summary of Professor Challenor's researches and describes the attempts (by the eminent philologist Sir James Murray) to check on this story. Needless to say, the result was that the legendary "Tan" and his seven books are completely unknown in China. Dudeney then goes on, apparently in all innocence, to contribute an additional piece of spurious mystification to the "tangram legend." He refers to a Chinese book of tangrams about which an American correspondent has written to him, and he reproduces a Chinese inscription from its first page. His reason for doing this is that—

> The owner of the book informs me that he has submitted it to a number of Chinamen in the United States and offered as much as a dollar for a translation. But they all steadfastly refused to read the words, offering the lame excuse that the inscription is Japanese. Natives of Japan, however, insist that it is Chinese. Is there something occult and esoteric about Tangrams, that it is so difficult to lift the veil?

* Two volumes of Loyd's puzzles have been republished recently, in which this propensity for fooling the reader is amply demonstrated. I refer to *Mathematical Puzzles of Sam Loyd* (two volumes), edited by Martin Gardner. New York: Dover Publications, Inc.

† Reprinted by Dover Publications Inc.

FIGURE 2

Now it so happens that I have a copy of this same book. The "mysterious inscription" is shown in Figure 2, along with the rest of the page on which it occurs. First of all we should remark that Dudeney, or his correspondent, has mistaken the back of the book for the front—a common error, since Chinese books start with what we would consider to be the last page. The inscription is actually on the last page but one. It is very badly printed, but otherwise there is nothing particularly difficult about it. One can only suppose that the people who were asked to translate it were less well versed in Chinese (or Japanese) than the owner of the book thought. The inscription reads roughly as follows: "Two men facing each other drinking. This shows the versatility of the seven-piece puzzle."

Clearly, this is simply a caption for the tangram outlines which appear beneath it. Far from being mysterious or occult, the inscription is not even out of the ordinary, since *every* outline in the book has a caption! What a shame! Another lovely theory ruined by hard facts!

Despite the popular proverb, truth is seldom stranger than fiction, and the full history of the tangram puzzle is probably not nearly as colorful as the legend that Sam Loyd and Dudeney have woven around it. But this in no way detracts from the enjoyment that we can derive from this pastime, as the following pages are intended to show. The object of the game is, of course, to put the seven tangram pieces together so as to form the various outlines given throughout the book. Some of these are quite easy, others less so. If an outline really stumps you, look up the solution at the back of the book. But the Solutions should be your last resort; don't consult them unless you are really stuck. You may want to try to construct some original tangrams, and some interesting possibilities are suggested here and there in the text.

Finally, a word about the name "tangram" itself. The "gram" part presents no difficulty; it is a common ending denoting something written or drawn (as in "diagram," for example). The origin of the first part of the word is more in doubt. There are several explanations, but one is so much more plausible than any other as to be almost certainly correct. The Tang dynasty was one of the greatest in Chinese history; so much so that in certain South Chinese dialects the word *tang* (or *tong*) is synonymous with "Chinese." Now, a European visiting China in the first half of the nineteenth century, if he learned any Chinese at all, would be more likely to learn a Southern dialect than any other. Thus a traveler bringing the puzzle from China to the West, and wanting a name for it, would be very likely to take the word *tang*—that is, Chinese—and combine it with the familiar ending "gram." Dudeney quotes Sir James Murray as being of the opinion that the name was probably introduced in this way by an American some time between 1847 and 1864, the latter date marking the word's first appearance in Webster's Dictionary.

If this conjecture about the derivation of the name tangram is correct, then the tangram is truly the "Chinese puzzle," in name as well as in origin.

1. *Letters and Numbers*

THE seven tangram pieces can be fitted together in such a vast number of ways that in preparing a selection of them it is very difficult to know where to start. Probably as good a way as any is to go right back to our early school days, back to the days when we first learned the alphabet. This time we shall make the letters out of the tangram pieces instead of with pencil and paper.

In the following pages you will find tangram outlines for all the letters of the alphabet. See if you can make them all yourself before looking at the solutions which begin on page 94.

Reconstructing tangram outlines that have been obtained by others is only half the fun that can be derived from this puzzle; as much entertainment, if not more, can be obtained by inventing new outlines. The reader can try his hand at this right away by trying to improve some of the outlines given for the letters. Certainly several could stand improvement; the N, Q and X, in particular, are nothing like as satisfactory as one would wish. The outlines given are the best that I have been able to concoct, but the reader may well be more successful.

For good measure, the digits 1, 2, up to 8 have also been given. Since zero is the same as the letter o, and 9 is simply a 6 upside down, these two digits are not new outlines, and have not been included.

1

2

3

4

5

6

7

8

9

10

11

12

13

14

15

16

17

18

19

20

22

21

23

24

25

26

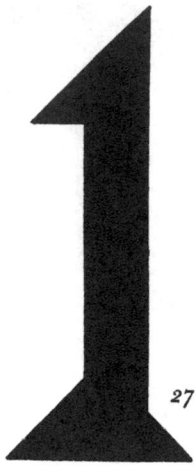

27

28

29

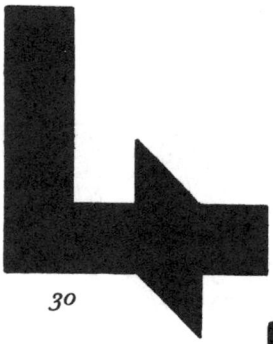

30

31

32

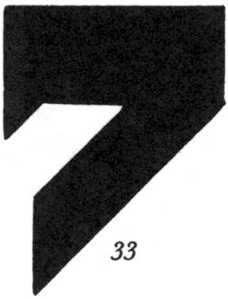

33

34

2. *Animals*

THE Chinese tangram books, and the later Western books all contain large numbers of outlines representing animals. This book will be no exception, and our contribution to the tangram menagerie lies in the next few pages. Some of the outlines which follow are taken, or adapted, from the Chinese books, and some from Sam Loyd's book. As far as I know, the others are entirely new.

It might be though that the tangram pieces, with their sharp angles and straight sides, could not possibly portray the complicated curved shapes of living creatures, but it is truly surprising how a well-designed tangram can suggest curves where none exist, and complexity where there is only simplicity. See, for instance, how a single piece suffices to suggest the back-curving horns of the mountain goat (No. 35); notice the stratagem by which the thin legs of the stork (No. 50) have been portrayed by three tangram pieces which, individually, are much thicker; note how the careful arrangement of the pieces unerringly conveys the fright of the startled cat (No. 41), the haughtiness of the camel (No. 47) and the graceful lines of the shark (No. 78). We shall see many examples throughout the book of outlines which, by their careful construction, convey much more to the mind's eye than is actually there.

In addition to the animals that have already been mentioned we have:

36. A horned animal of some kind. (Let's not inquire too closely into its precise genus and species!)

37. A polar bear.

38. A giraffe.

39 to 44. Assorted cats, closely followed (naturally) by

45 and 46. Two dogs.

48. A squirrel.

49 to 61. Birds of various shapes and sizes.

62 to 65. Four sinister-looking vultures.

66 to 68. Three horses. Also No. 74.

69 to 72. Four different kinds of bats.

73. A kangaroo.

75. A crocodile.

There follow several sea creatures, mainly fish of various kinds, but also

79. A lobster.

81. A turtle.

82. A seal.

83. A shrimp.

THE TANGRAM ZOO

35

36

GOATS

Polar Bear

37

38 *Giraffe*

40

39

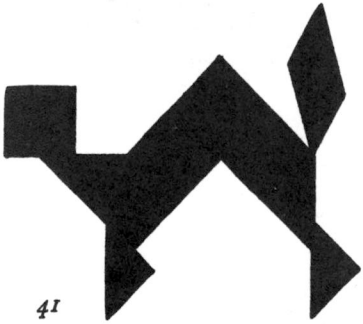

41

CATS

CATS AND DOGS

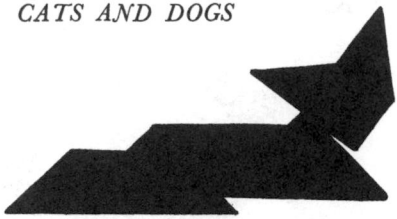

42

43

44

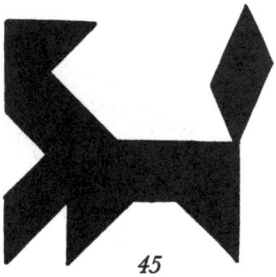

45

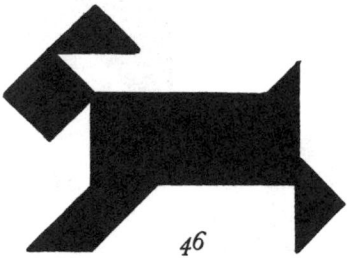

46

47 Camel

48 Squirrel

BIRDS

MORE BIRDS

62

63

VULTURES

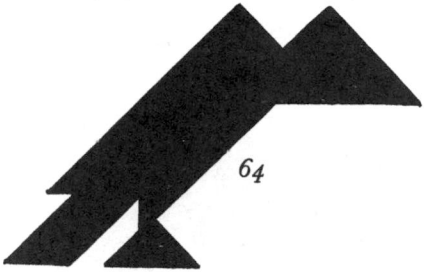

64

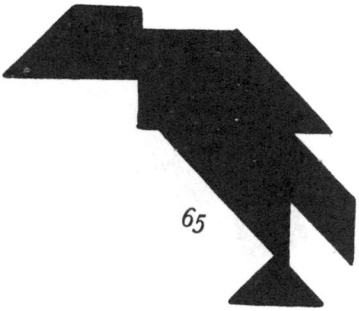

65

66

67

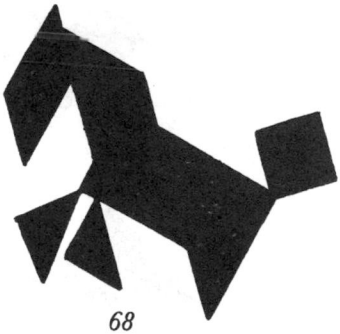

68

HORSES

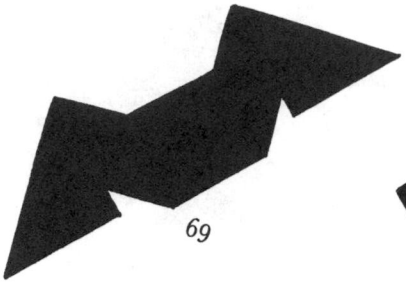

69

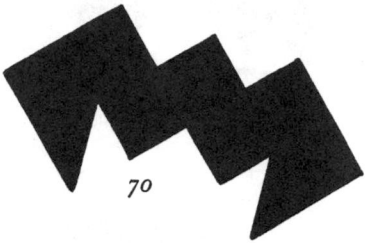

70

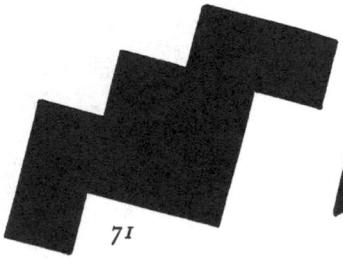

71

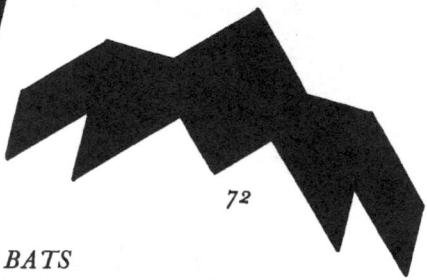

72

BATS

73 *Kangaroo*

74 *Horse*

75 *Crocodile*

SEA CREATURES

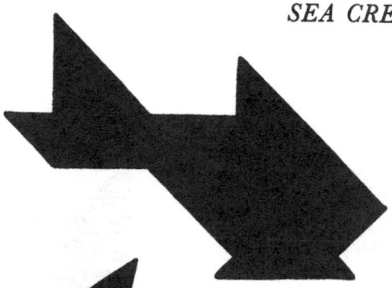

76

77

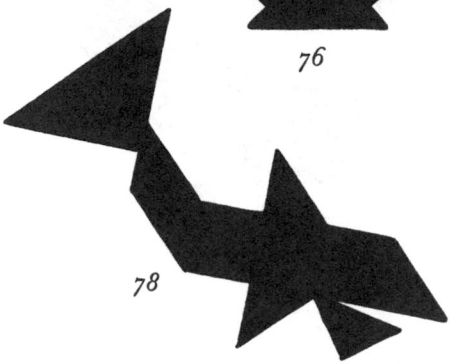

78

79 Lobster

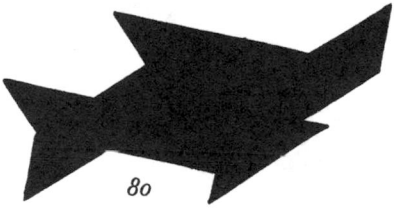

80

81 Turtle

82 Seal

83 Shrimp

SEA CREATURES

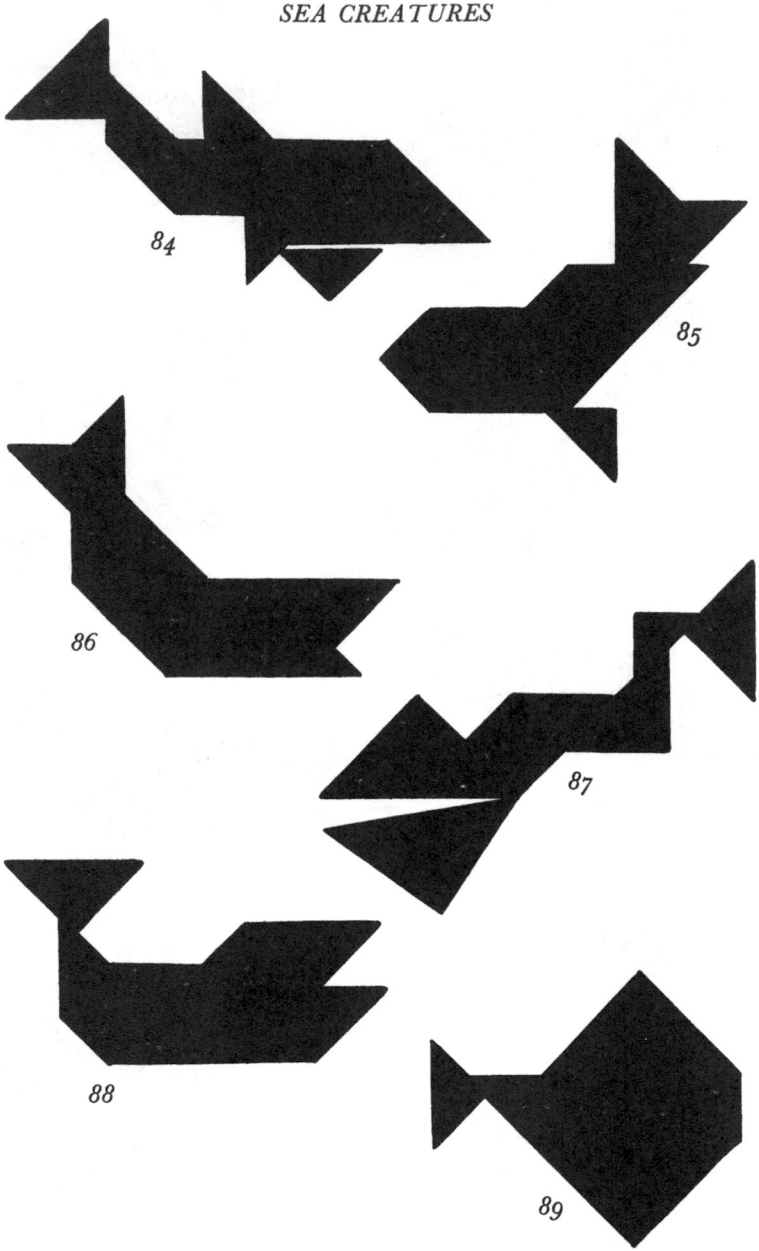

3. *Mankind*

WHILE the evolutionist will not deny the propriety of dealing first with animals, we should delay no longer in looking at those tangrams which depict man and his activities. These are among the most interesting of tangrams: for since the human shape is more familiar to us than that of any other object in the universe, we can readily appreciate a tangram which succeeds in portraying the essence of some human form or posture; and just as readily detect the fault in those that do not manage to suggest what they are meant to represent.

Our human parade starts with a shoeshine boy and a customer (Nos. 90 and 91), two men very excited about something, and three acrobats (94, 95, 96), one of whom has just had a fall (or perhaps he is just taking a rest). On the next page are five men who are obviously in a great hurry and a sixth, strolling along with his hands in his pockets, who is clearly in no rush to get anywhere (97 to 102). Then we have four stately medieval ladies (103 to 106) accompanied by a servant (107) and a lady's maid (108).

Next we have "Sam Loyd's Portrait Gallery," twenty-six cleverly constructed heads taken from Sam Loyd's book, plus two more given by Dudeney. Sam Loyd gave names to some of these outlines, as follows:

120. Old Scotch Piper.
121. French Grenadier.
122. Colonial General.
123. A Turk.
124. Aunt Betsy.
125. Uncle Rhube.
126. Mary Smith.
128. John Knox.

129. Tom Sharkey.
130. The Professor.
131. Buffalo Bill.
132. "The Easy Boss."

The Indian chief and his squaw (137 and 138) are also from Sam Loyd's book.

Finally we have four horsemen (139 to 142), a runner (143), and a lady and a gentleman drinking a toast (144 and 145). (Or is he a gentleman? He seems to be seated while the lady is standing—and seated on the floor at that!)

90 91

Shoeshine Boy and Customer

92

93

94

ACROBATS

95

96

97

98

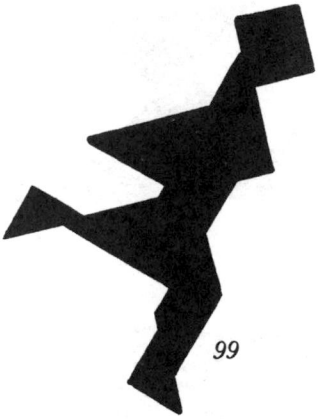

99

100

101

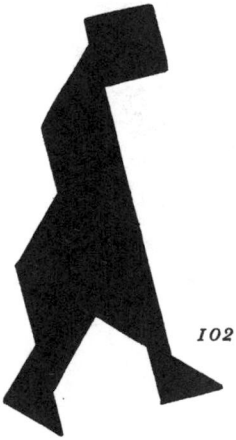

102

MEDIEVAL LADIES

103

104

105

106

AND SERVANTS

107

108

109

110

111

112

113

114

115

116

117

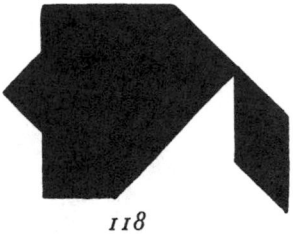

118

119

120

121

122

123

124

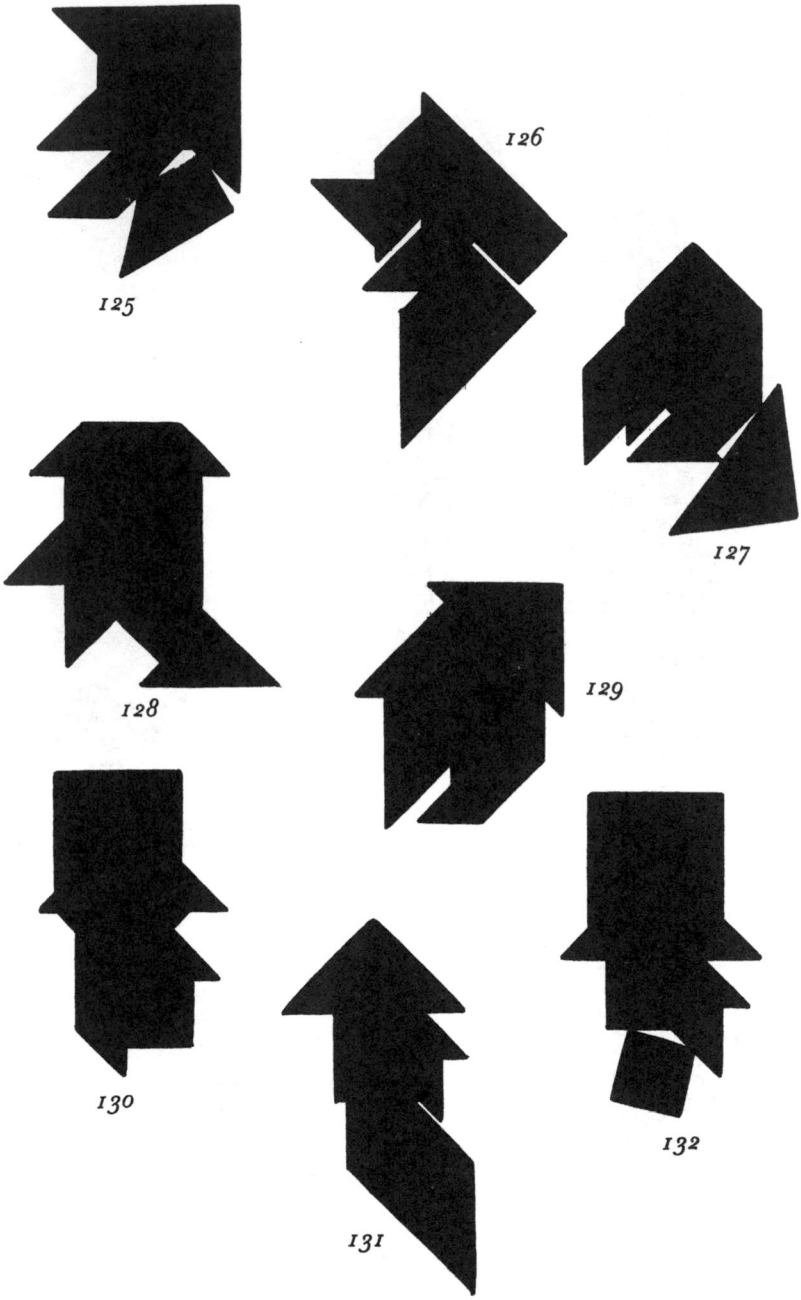
125
126
127
128
129
130
131
132

133

134

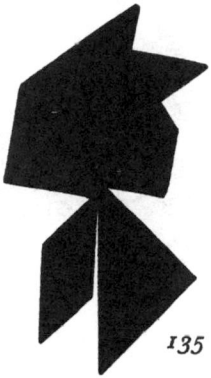

135

136

137

138

HORSEMEN

139

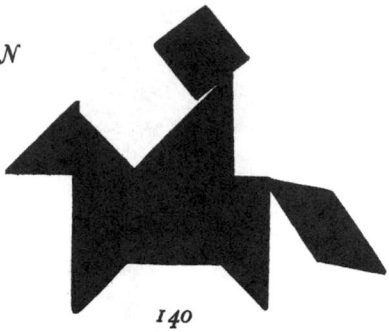

140

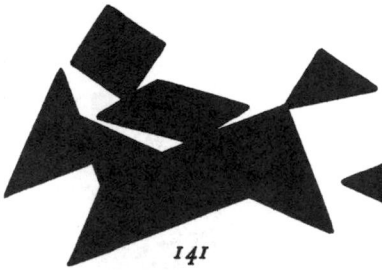

141

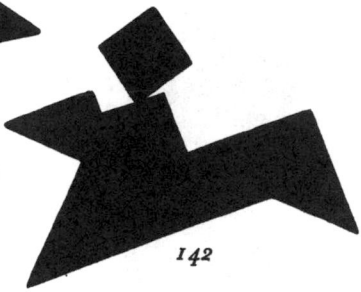

142

143

144

145

4. *Around the House*

It is only fitting that after looking at man and his activities we should dwell for a few moments on the things that he uses every day. Starting with his pipe (No. 146) we go through a motley collection of shoes, chairs and general odds and ends, ending up with a watering can and a pistol (Nos. 161 and 162). We then have three different kinds of baby carriages—all the latest models and very comfortable, despite the fact that they have square wheels (to which tangram babies have long grown accustomed!). Number 165 is a very superior model which comes complete with a nursemaid (No. 166) to push it. Finally we have Number 167, which is a—now what on earth was that thing meant to be? A saw? A nutmeg-grater? Well, no matter; it is an interesting shape to construct.

146 Pipe

147

148 Candle

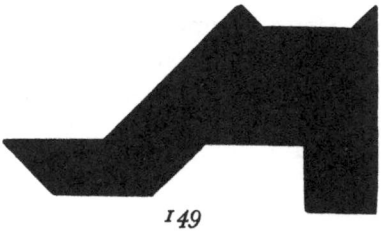

149

150

151

152

SHOES

153

154

CHAIRS

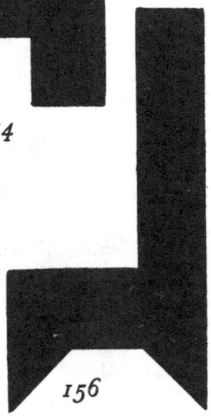

156

155

157

158 *Shirt*

159 *Stocking*

160

161 Watering can

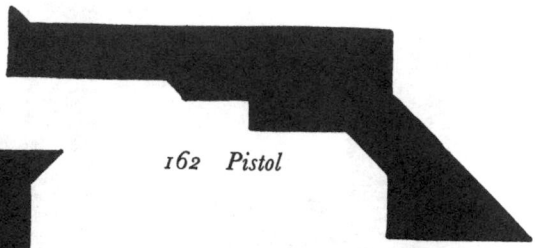

162 Pistol

163

164

BABY CARRIAGES

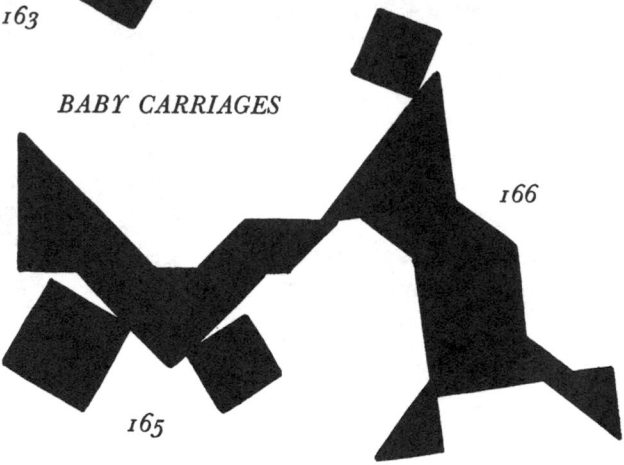

165

166

167 ???

5. *Boats and Bridges*

THE Chinese books which served to introduce the tangram to the Western world were published in the Treaty Ports of South China, and their pages are full of representations of ships, boats, junks and many other things connected with the sea. It is a curious fact that sailing boats of various kinds seem to be comparatively easy to depict using the seven tangram pieces, notwithstanding the rather complicated silhouettes which they present; whereas steamships, which with their more angular shapes might seem much more suitable for representation by tangrams, are in fact very difficult to concoct. I have been quite unable to produce a convincing tangram of any sort of mechanically propelled craft, with the exception of the launch given in Number 186 (and I won't argue with the reader if he says that this tangram, too, is not very convincing).

Here, then, is something on which the reader can exercise his creative ability; to produce a reasonably good outline of some kind of steamboat or other modern sea-going craft. It will not be easy!

Along with the ships and boats is given an assortment of bridges, mostly taken from the Chinese books. The last two tangrams of this chapter show a lighthouse and the lighthouse keeper being rowed out to it. (Or maybe it is just a sailor taking his maiden aunt for a row around the bay.)

BOATS

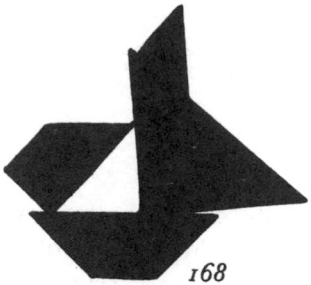

168

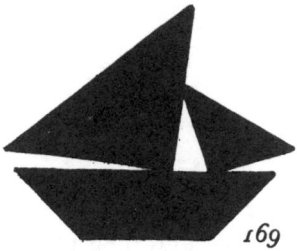

169

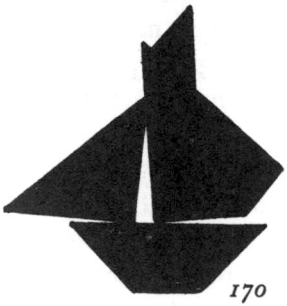

170

171

172

173

AND BRIDGES

174

175

BRIDGES

176

177

178

179

180

181

182

BOATS

183

184

185

186

187 & 188 Rowing to the lighthouse

6. *Stories and Pictures*

Two innovations introduced by Sam Loyd and H. E. Dudeney to extend the scope and interest of the tangram were to construct a series of outlines to illustrate a story, and to bring together several tangram outlines to make a more detailed picture. In this chapter we have two illustrated stories by Sam Loyd, and two pictures by Dudeney.

The first set of tangrams illustrates the well-known nursery rhyme of "The House That Jack Built." In outlines 189 to 199 we see

> . . . the farmer sowing his corn,
> That kept the cock that crowed in the morn,
> That waked the priest all shaven and shorn,
> That married the man all tattered and torn,
> That kissed the maiden all forlorn,
> That milked the cow with the crumpled horn,
> That tossed the dog,
> That worried the cat,
> That killed the rat,
> That ate the malt,
> That lay in the house that Jack built.

To fill out the page, three outlines are given, representing animals that would very likely be found in the vicinity of "The House That Jack Built."

The second story (adapted from Sam Loyd's version) is that of Cinderella. We see Cinderella crying in front of the fireplace (204 and 203); the two ugly sisters (205 and 206), and the fairy godmother with the pumpkin that became a coach and two of the rats that became coachmen (207 to 210). Cinderella dances with the prince (211 and 212), heedless of the clock (213) which is about to strike twelve. The episode of the slipper (214) is too well known to need repetition; everyone knows the story ends with wedding bells (215 and 216) for Cinderella and her prince.

One outline that clearly ought to have been included in this sequence is Cinderella's coach (when it wasn't being a pumpkin). Sam Loyd does not give one in his book, and I have been unable to devise a satisfactory tangram coach; they all turn out looking like baby carriages! Here again the reader may be more successful. By way of compensation I have included Cinderella's coach among the double tangrams later on in this book.

On the next page we have a picture by H. E. Dudeney entitled "A Game of Billiards." Dudeney says, "The players are considering a very delicate stroke at the top of the table."

Dudeney's second picture consists of nine outlines (221 to 229) and will not fit on a single page of this book, so has been spread over two. The description of it is best left to Dudeney himself.

> My second picture is named "The Orchestra", and was designed for the decoration of a large hall of music. Here we have the conductor, the pianist, the fat little cornet-player, the left-handed player of the double-bass, whose attitude is life-like, though he does stand at an unusual distance from his instrument, and the drummer-boy, with his imposing music-stand. The dog at the back of the pianoforte is not howling; he is an appreciative listener.

THE HOUSE THAT JACK BUILT

Illustrated by Sam Loyd

189

190

191

192

193

194

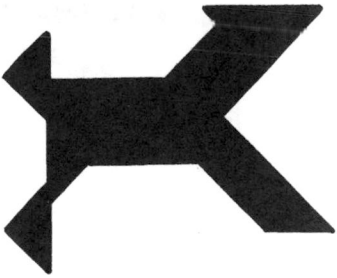

195

196

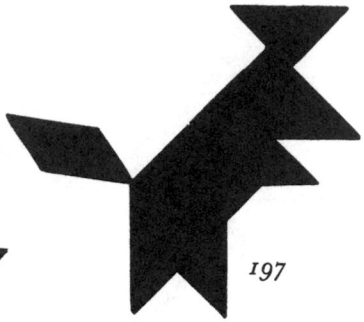

197

198

199

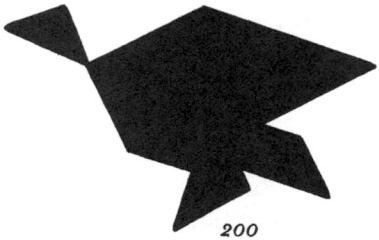

200

201

202

THE STORY OF CINDERELLA Illustrated by Sam Loyd

203

204

205

206

207

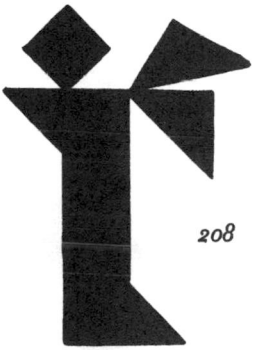

208

209

210

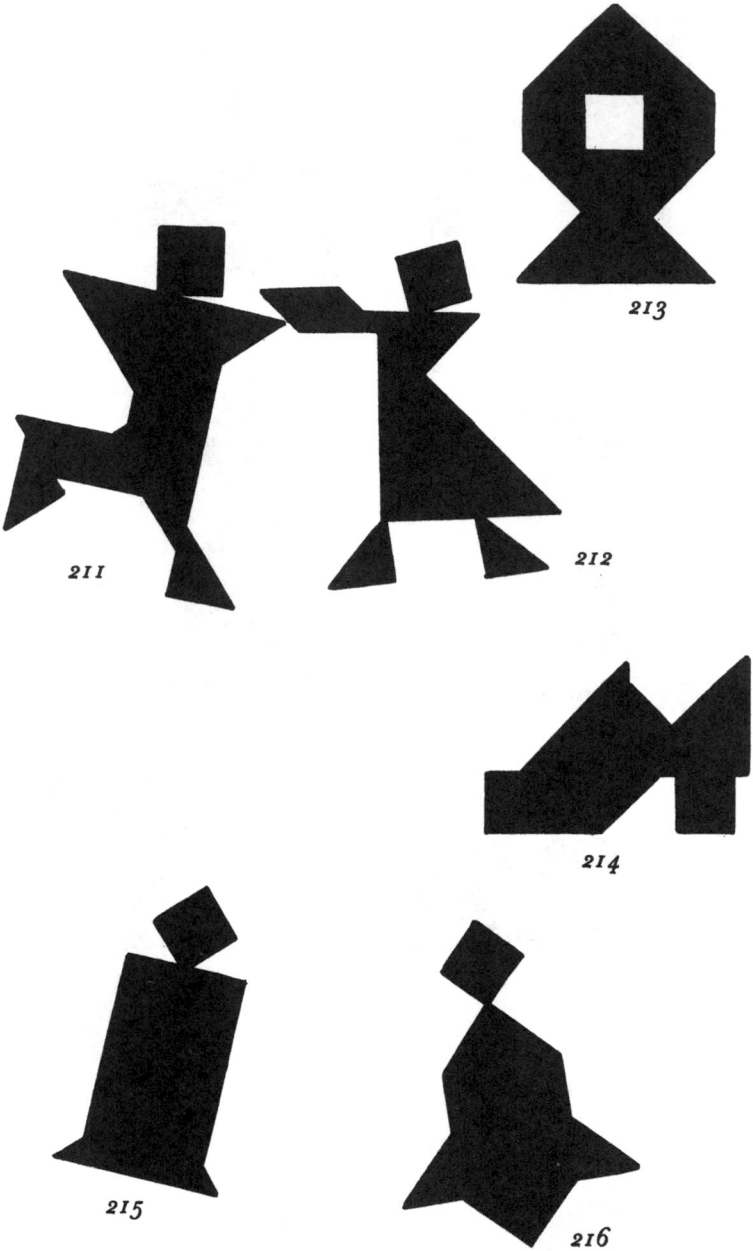
211
212
213
214
215
216

A GAME OF BILLIARDS

By H. E. Dudeney

217

218

219

220

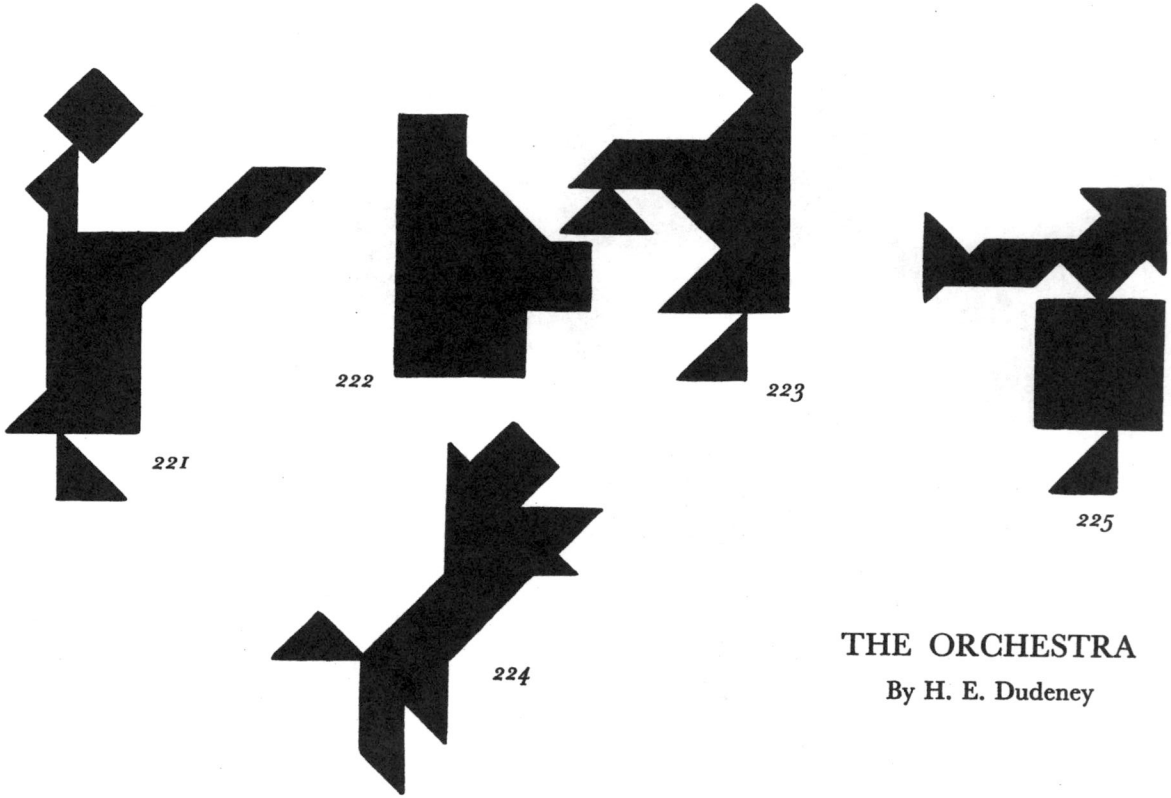

THE ORCHESTRA

By H. E. Dudeney

229
228
227
226

7. *A Little Mathematics*

(DON'T be alarmed! You will need very little knowledge of mathematics to follow this chapter.)

As we have already remarked more than once, the number of tangrams is very large. This is really an understatement, for the number is infinite! This we can easily see by looking at Number 229, for example. The bottom corner of the square piece that represents the head of the drummer-boy can touch the rest of the tangram at an unlimited number of points along the line that represents the shoulder and the outstretched arm. It is true that the different outlines that one would get by putting this piece in the different positions would not be *very* different each from the other, but in the strictest sense they would have to be counted as distinct.

It is a little quixotic, however, to accord the same weight to minute variations in a tangram outline as to the differences between two totally distinct outlines, and one naturally wonders whether, by ignoring trivial variations, one could ask the question, "How many tangrams are there?" with some hope of getting a clear-cut answer. Alternatively one might ask the question, "How many tangrams are there of such-and-such a particular kind."

In 1942, two Chinese mathematicians, Fu Tsiang Wang and Chuan-Chih Hsiung, asked, and answered, the question, "How many convex tangrams are there?" Now, before we go further we must be sure what is meant by "convex" in this connection. Roughly speaking, we can say that a convex figure is one that does not have any recesses in its outline. For angular figures (like tangram outlines) this means that all the angles are less than 180°; in other words that the corners all stick out instead of in. But the simplest way of seeing the difference between a convex figure and one that is not convex is to imagine a piece of string or an elastic band pulled tight around the figure, as in Nos. 230 and 231, below. If this causes the string to make contact with the figure all the way round its edge, then the figure is convex; but if there are gaps

230

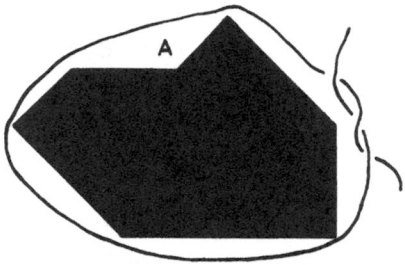

231

between the string and the edge of the figure, as at A in Number 231, then the figure is not convex, but concave or recessed, where these gaps occur.

Now that we know exactly what "convex" means in this context we can look again at the question asked by the Chinese mathematicians: "How many convex tangrams are there?" One might well imagine that there would be quite a large number of them, but it turns out that there are only thirteen! In the paper in which this result is proved* the thirteen convex tangrams are not drawn, but are merely listed as follows:

Triangles	1
Four-sided figures	6
Five-sided figures	3
Six-sided figures	3

In this chapter we give all thirteen convex tangrams. Some of them are very easy to construct, but a few are rather tricky. The reader may want to try to construct them all without first looking at the outlines given below (Nos. 232 to

* Published in the *American Mathematical Monthly*. Vol. 49 (1942), p. 596.

243). Or he can examine the outlines and then try to construct the tangrams. The total of thirteen is completed with Number 230, but since this tangram was necessarily given in the introduction, it doesn't really count.

An Unsolved Problem

It is clear that convex tangrams are very special indeed; there are so few of them. Can we think of any other special kinds of tangrams that would be more numerous than the convex ones, and yet not infinite in number? As far as I know, no one has ever done so, but I am going to propose a problem of this kind for the benefit of anyone who may feel inclined to tackle it. First we must specify the sort of tangram that we are going to talk about.

Let us imagine a set of tangram pieces of such a size that the equal sides of the small triangles are 1 inch in length. Then the third side of these triangles will be of approximately 1.414 inches (the square root of 2, to be precise). Now any side of any of the pieces of this set will be one of these lengths, or twice one of these lengths, and we can therefore imagine every side of each of the pieces to be made up of "sections" whose lengths are either 1 inch or 1.414 inches. There will be either one or two sections to each side. In Figure 3 (page 58), which shows the tangram pieces, the ends of the sections are indicated by blobs.

Imagine now a tangram that has been constructed in such a way that wherever two pieces are in contact at all, they are in contact along a whole section of each, so that the ends of these sections coincide. In other words, when two pieces are in contact, the blobs on the two edges will match. This is illustrated by Figure 4. This stipulation on the way in which the pieces are to be placed together considerably restricts the sort of outline that can be produced; on the other hand, large numbers of very interesting tangrams fall into this category. We shall apply one further restriction: namely, that the tangram should be all in one piece. Tangrams which conform to the above restrictions I call "snug" tangrams, because of the close way in which the pieces fit together. All the convex tangrams are snug; so is Number 231, and so are very many other tangrams in this book. Snug tangrams tend to be rather more difficult to reconstruct from their outlines than tangrams that are not snug, since the close fitting reveals less of the way in which the tangrams have been formed.

It makes sense to ask the question, "How many snug tangrams are there?" for it can be shown that snug tangrams, unlike tangrams in general, are limited in number—there is only a finite number of them. But how many, exactly? At the moment, nobody knows, and I recommend the problem of calculating this number to anyone who finds it interesting and who has access to a large electronic computer. It is unlikely that the "snug tangram number" will be found without the use of a computer, for it is almost certainly very large, probably well up into the millions, if not considerably larger.

(A preliminary investigation of the problem indicates that it is too complex for the computer to which I have access, but that the problem of programming a computer to find the number of snug tangrams would be a fascinating one—a puzzle that out-tangrams the tangram! Unfortunately, it is a puzzle not available to everyone.)

232

233

234

235

236

237

238

239

240

241

242

243

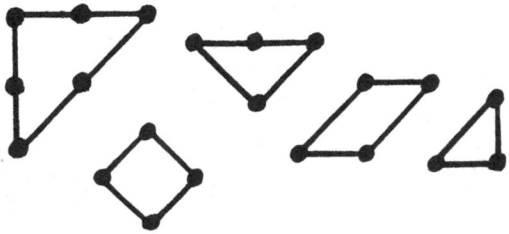

FIGURE 3

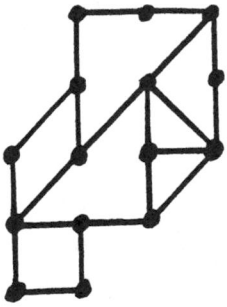

FIGURE 4

8. *Chinoiserie*

As we remarked in the Introduction, the first books of tangrams available in the West were exact reprints of Chinese books on the subject. Some of the objects depicted in these books—ships, birds, and so on—were familiar to Western readers, but the majority of the outlines represented things essentially Chinese, having no Western counterparts. To the editors in England and America these outlines were just so many arbitrary geometrical shapes, and they were treated as such. Yet many of them represent objects of considerable interest, either in themselves or because of their connection with outstanding events in Chinese mythology, literature or history. For this reason we shall look at a few of them in this chapter.

We started this book by forming the letters of the alphabet. The Chinese books did a rather similar thing, but with one important difference. The Chinese schoolboy—pity him!—does not have a paltry twenty-six symbols to learn, as we did, but several thousand characters, many of them very complicated in shape. (To get an idea of their complexity you need only look at our frontispiece, which reproduces the title page of a Chinese tangram book.) Most Chinese characters are far too intricate to be represented in tangram outline, but it is possible to represent some of the simpler ones, and these were not overlooked by the Chinese tangram-puzzlers.

Numbers 244 to 250 give the Chinese characters for "big," "small," "six," "under," "above," "of," and "mountain." (The characters as usually written are: 大, 小, 六, 下, 上, 之 and 山.) The remaining outlines of this chapter are listed below, with comments where necessary.

251. A goldfish. Goldfish came originally from China and are still extremely popular there, as elsewhere.

252. This is called "riding a crane." The ancient

Chinese sages, unlike present-day philosophers, had no desire to go traveling about all over the world, and were quite content to stay in one place to meditate. However, when the need did arise to travel from one place to another, they would do so on the backs of cranes. At least, this is what was popularly supposed; presumably, the cranes of those days were stronger birds than they are now!

The tangram represents one of these immortals mounted on his strange steed. The upper parts of the crane and the rider can be seen, but the nether regions are apparently left to the imagination.

253. A bottle gourd.

254. This is a kind of citrus fruit known as "Buddha's hand."

255. A helmet of a kind once worn by Chinese soldiers.

256. This is entitled "Drinking Alone," and alludes to a very well-known poem by the Tang dynasty poet Li Po, "Drinking Alone by Moonlight." The lonely poet invites the moon and his shadow to join him in his solitary tippling.

257. A clock—probably of Western manufacture.

258. A brazier.

259. A dagger.

260. A pinking knife. (Compare the dressmaker's pinking shears.)

261. A teapot—a familiar enough object to us, too.

262. A mulberry hook—presumably for cutting or trimming mulberry bushes and trees. These would be important to the Chinese, since mulberry leaves constitute the staple of the silkworm's diet.

263. This is a raincoat with a hood. As a tangram outline it shows a clever use of empty space.

264. A soup spoon. Any reader who has had a meal at a Chinese restaurant will know the sort of thing that is meant. It is made of china instead of metal, as our spoons are.

265. Another teapot.

266. A Dragon Boat. This refers to a very old Chinese festival, the Dragon Boat Festival, which was held on the fifth

day of the fifth month. Everyone repaired to the river banks, where boats gaily decked out to resemble floating dragons raced up and down the river. The tangram represents one of these boats.

267. The Three-legged Toad. In China, as elsewhere, there are many stories and legends about the moon, and these stories often center around some interpretation of the markings that can be seen on the moon's disk. According to one such legend the markings represent a three-legged toad that lives in the moon. (Rather like our "Man in the Moon.") The tangram is an attempt (not too successful, I fear) to depict this strange animal.

268. A hexagonal pendant. Another tangram that shows good use of empty space.

269. A carp.

270. A flower bowl.

271. A pair of scissors.

272. A saw. (No handles, apparently!)

273. A clothes post. Two of these were used to support a thin bamboo pole on which clothing was hung to dry. This same tangram is elsewhere described as an incense table.

274. A bell.

275. A feather tube. The ceremonial dress of Chinese generals used to include several very long feathers. (These are often very conspicuous in pictures of scenes from Chinese operas, in which generals frequently appear.) When not in use, these feathers were kept in special tubes in order to prevent their being damaged. In much the same way women on both sides of the Atlantic, around the turn of the century, protected the ostrich feathers they wore in their hats.

276. A lucky charm.

277. A look-out tower.

278. A medicine jar.

279. A rack for a hand mirror.

280. A ladle.

281 and 282. Two baskets.

283. This is a stylized arrow with a flag attached, which the Emperor conferred as a symbol of authority.

284. A "foreign candle"; i.e., one imported from the West.

285. A water pot.

286. A measure.

287. A mallet.

288. A Western knife; i.e., a penknife.

SOME CHINESE CHARACTERS

244

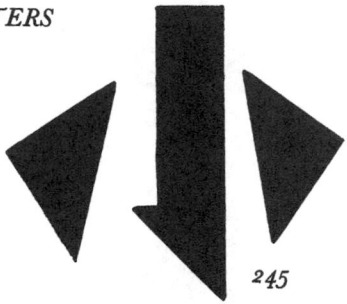

245

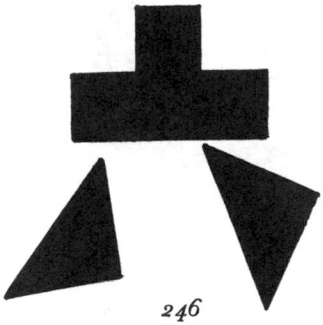

246

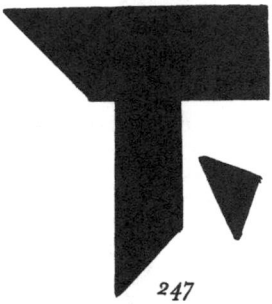

247

248

249

250

251 Goldfish

252 Riding a crane

253 Bottle gourd

254 "Buddha's hand"

255 Helmet

256 "Drinking alone"

257 Clock

258 Brazier

259 Dagger

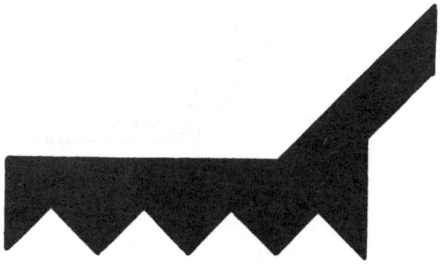

260 Pinking knife

261 Teapot

262 Mulberry hook

263 Raincoat

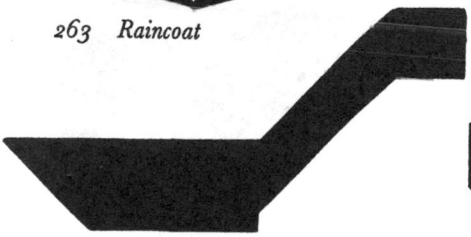

264 Soup spoon

265 Teapot

266 Dragon boat

267 Three-legged toad

268 Pendant

269 Carp

270 Flower bowl

271 Scissors

272 Saw

273 Clothes post

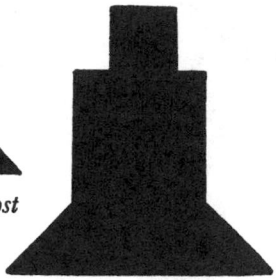

274 Bell

275 Feather tube

276 Lucky charm

277 Look-out tower

278 Medicine jar

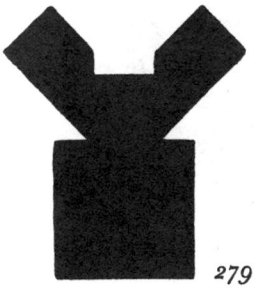

279 Mirror rack

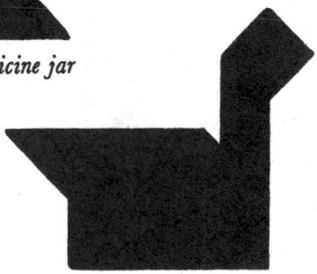

280 Ladle

TWO BASKETS

281

282

283
Arrow

284 Candle

285 Water pot

287 Mallet

286 Measure

288 Penknife

9. *Paradoxes and Illusions*

BOTH Sam Loyd and H. E. Dudeney discovered that the tangram could be the source of some pretty paradoxes. The Chinese puzzlers, it would appear, never hit on this particular aspect of the puzzle. Indeed, their diagrams are so carelessly given (many of them being quite wrong!) that it would have been very difficult for the readers of these books to appreciate the subtle differences between tangram outlines on which these paradoxes depend.

Let us start with a paradox by Dudeney. In Numbers 289 and 290 we have two stately gentlemen who appear to be identical in all respects except that one has a foot while the other has not. Yet each of these tangrams uses all seven of the pieces. Whence, then, does the second gentleman get his foot?

This paradox is similar to, and just as effective as, a paradox that has long been known to mathematicians. Take a square whose sides are, say, 8 inches in length, and divide it into four parts as shown in Figure 5, below. These four pieces

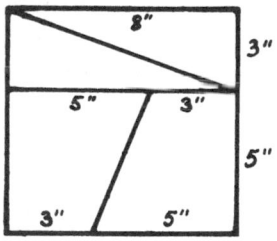

FIGURE 5

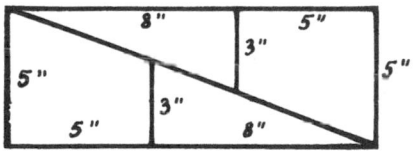

FIGURE 6

can then be rearranged, as in Figure 6, to form a rectangle measuring 5 inches by 13 inches. At once we see the paradox;

for the area of the original square was $8 \times 8 = 64$ square inches, while the area of the rectangle is $5 \times 13 = 65$ square inches. Where has the extra square inch come from?

If the reader does not believe that the four pieces can be fitted together as shown, I suggest he put it to the test by cutting up a square in the manner indicated, and trying the rearrangement himself. An account of this paradox, together with an explanation of it, and many other paradoxes of a similar kind are to be found in *Mathematics, Magic and Mystery* by Martin Gardener.*

There are many other tangram paradoxes of this kind. Numbers 291 and 292 represent two flatirons, each made with all seven tangram pieces. Yet one of them has an extra knob! Again, in Numbers 293 and 294 we have two snuffboxes, and the knob on the second one is twice the size of the knob on the first!

Then we have four vases (295 to 298), all apparently of the same size. Yet only the last is complete. Number 295 has a chip out of the rim, while 296 and 297 each have a hole in the middle, but holes of different sizes! Tangrams 299 and 300 are not very different in outline, but it requires two entirely different arrangements of the pieces to make them.

The reader who has sorted out the difference between Tangrams 289 and 290 (or who has peeped at the Solutions at the end of the book) should have no difficulty in constructing these tangrams, and explaining the paradoxes that they present.

A different sort of paradox, or perhaps "illusion" would be a better word, was invented by Sam Loyd, and is illustrated by the next eight tangrams. They are those used by Loyd in his book. Number 301 is an ordinary wrench; 302 is a rather more elaborate kind of wrench, with a knob to adjust it. The handle is impossibly short, but let's not be too fussy.

Tangram 303 is a Japanese girl in traditional costume, walking gracefully across the floor; 304 is the same Japanese girl executing a polite bow, displaying the bustle which is a characteristic part of the traditional dress. Tangram 305 is yet

* New York: Dover Publications, Inc.

another baby carriage, and 306 a version with large wheels in front.

The reader should try to construct these six tangrams before reading any further.

After the reader has constructed these tangrams he should answer (honestly) the following question: Did you notice that Tangrams 302, 304 and 306 are identical, except for the position in which they are drawn?

A similar illusion is presented in Tangrams 307 and 308, showing a man pushing a wheelbarrow. One could very easily overlook the fact that these two tangrams are exactly the same!

We return to the original type of paradox in Tangrams 309 and 310. One of these pigs has lost his tail. Where did it go?

Finally, from the tangram pieces, which were originally cut from a complete square, we are asked to make a square with a square corner removed from it. The side of the piece removed is one-third the length of the side of the big square. Despite the apparent paradox, this can be done. What is more, the seven tangram pieces can be divided into three groups in such a way that the transition from the complete square to the square-with-a-bite-out-of-one-corner can be made simply by moving these groups *en bloc*. Solve this tangram puzzle and you will have solved the following dissection problem: With two straight cuts, divide a square into three parts which can be rearranged to form a square with a square portion (of dimensions given above) missing from one corner. These two cuts are indicated by extra heavy lines in the Solutions section at the end of this book.

289

290

291

292

293

294

295

296

297

298

299

300

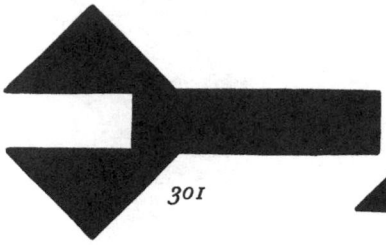

301

302

303

304

305

306

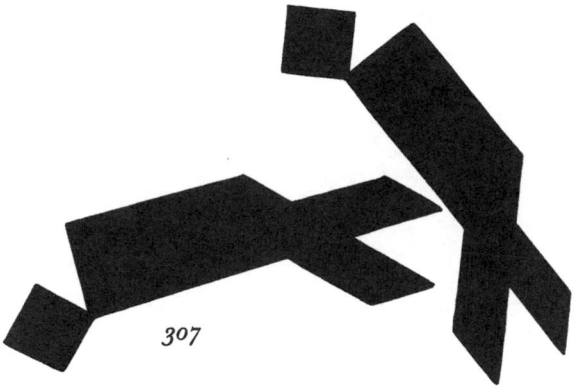

307

308

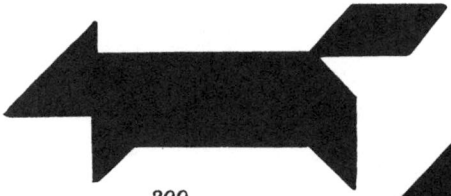

309

310

311

10. *Double Tangrams*

TANGRAMS are amazingly varied and numerous, but there are, of course, limits to what can be done with only seven pieces. It is natural, therefore, to try the effect of using two or more sets of tangrams in an endeavor to achieve more intricate results. Dudeney's pictures (Chapter 6), even though they use several tangram sets, are nevertheless in the true tangram tradition, since each element in the picture is a single complete tangram. In this chapter we shall look at some "double tangrams," made from two sets of tangram pieces, in which the fourteen pieces are used all together, with no separation between the pieces of one set and those of the other.

As the reader will see from the outlines given in this chapter, much more intricate shapes are possible with a double tangram set; yet, on the whole, these double tangrams are less interesting than their single brethren. With fourteen pieces to play around with, one cannot help but feel that it should be possible to arrive at a reasonable likeness to just about anything. Consequently, the sense of achievement that one gets on producing a recognizable cow, sailing boat, human figure, or what have you, from a mere seven pieces, is quite lacking. For all that, the double tangrams are not altogether without interest, and certainly deserve mention. In the following pages we give a selection of double tangram outlines.

Numbers 312 to 317 are self-explanatory (I hope). Number 318 is Cinderella's coach, which we promised in Chapter 6. (If you think it still looks like a baby carriage, I can only say, "Go and make a better one, then!") Then we have a space rocket and a television set (We're really up to date now!), followed by a locomotive. Finally, a microscope (322), a telescope (323), and that most fiendish of all inventions of the devil, an alarm clock!

312 House

313 Cyclist

314 Dog

315 Lion

316 Car (ancient vintage)

317 Sports car

318 Coach

319 Rocket

320 Television set

321 Locomotive

322 Microscope

323 Telescope

324 Alarm clock

11. *The Fifteen-Piece Puzzle*

In 1962, while searching in the British Museum for further information on tangrams I came across two volumes printed in China in the latter half of the nineteenth century dealing with a puzzle which will be of interest to those who find the tangram interesting. These books were written by one T'ung Hsieh-keng, and the Chinese name for the puzzle is literally "Increase-knowledge puzzle"; we shall call it the fifteen-piece puzzle. As this name suggests, fifteen pieces are used, and they are cut from a square in accordance with the following figure:

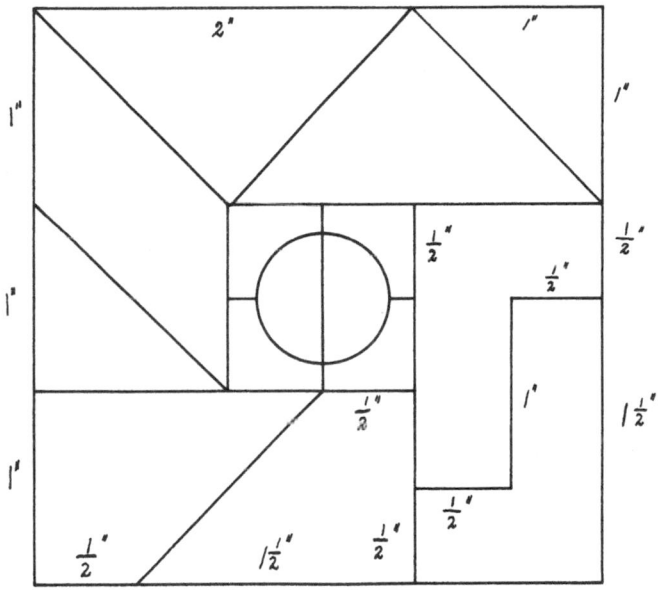

Figure 7

Suggested measurements are given, though the square can be of any size. With these measurements the diameter of the central circle is $\frac{2}{3}$ inch. As with the tangram pieces, all the angles are either 45° or 90° or 135°.

What was said in the last chapter about the lesser interest of the double tangrams, compared with single tangrams, will apply equally well here; with fifteen pieces one ought to be able to make a reasonable outline of almost anything. The fifteen-piece puzzle has some added points of interest, however, in that the pieces are more varied in shape; there are even some curved edges.

The two Chinese volumes contain a large number of pictures made up from the fifteen pieces. Many of them are straightforward outlines of single objects, not so very different from the tangrams or double tangrams given in this book; but many of them are complete pictures, very much in the traditional style of Chinese painting, and with a typically Chinese choice of subject matter—a pavilion in the moonlight, a man asleep in a drifting boat, and so on. As is typical in Chinese paintings, an appropriate poetic quotation is written in one corner of the picture.

Unlike the diagrams in the Chinese tangram books of the early nineteenth century, the illustrations in these two volumes are carefully and pleasingly drawn. For this reason the outlines appearing in the next few pages are given exactly as in the original. They are representative of the many delightful constructions that these two books contain.

The reader who has derived pleasure from trying to make the tangrams in the first ten chapters will surely want to cut himself a set of fifteen pieces and endeavor to reproduce these pictures. In some of the solutions of these puzzles given at the back of the book the relative positions of the various parts of the picture have been altered for the sake of convenience; the construction of the individual parts, however, is correctly given.

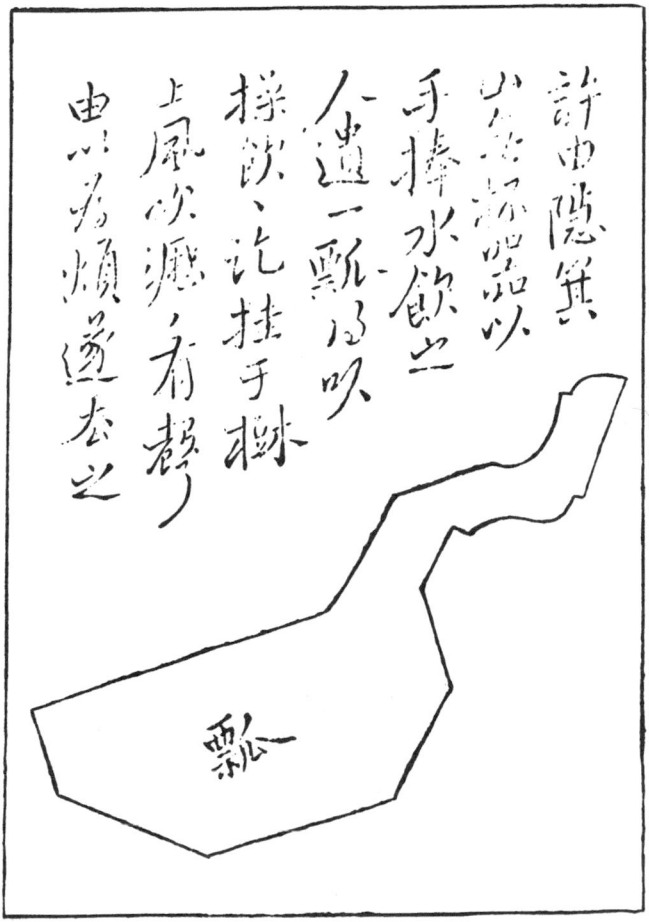
許由隱箕
山之下洗耳以
手捧水飲之
人遺一瓢可以
採飲之乃挂于樹
上風吹瀝有聲
由以為煩遂去之
瓢一

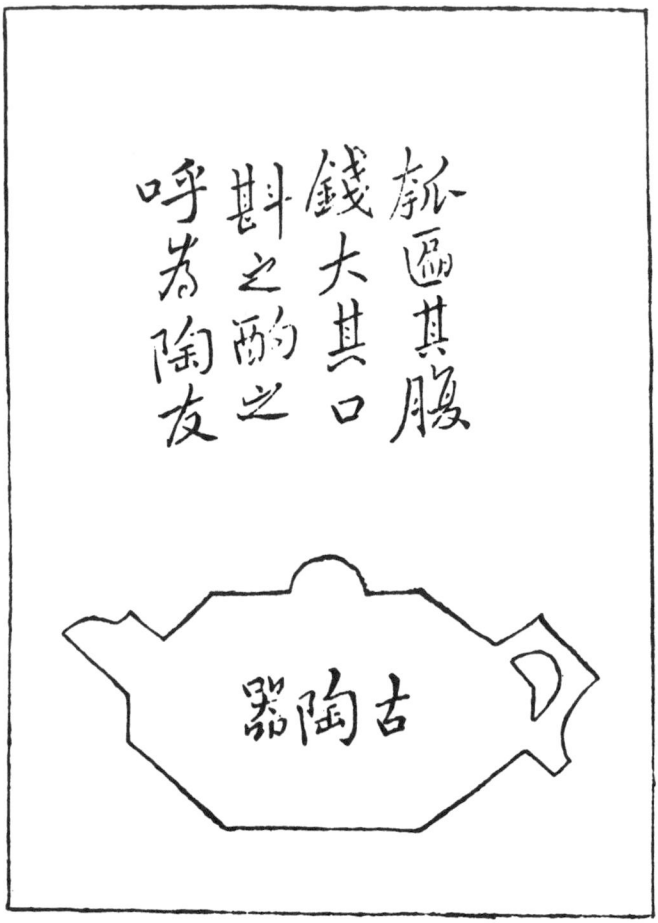

326

不炊而沸
不汲自盈
絪緼之氣
自然所生

鼎三足
兩耳
和五
味之
瑤器也

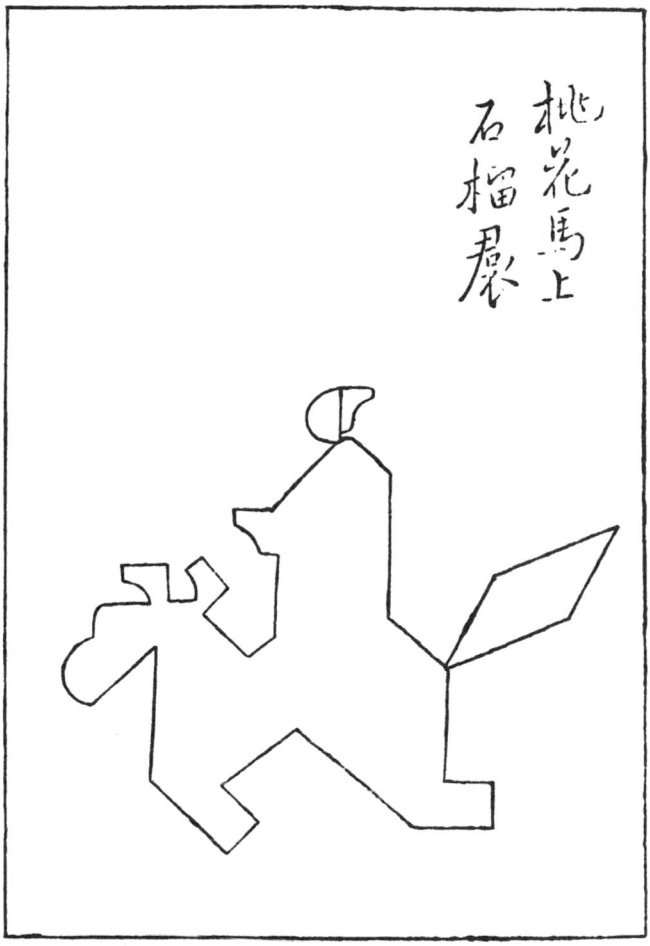

329

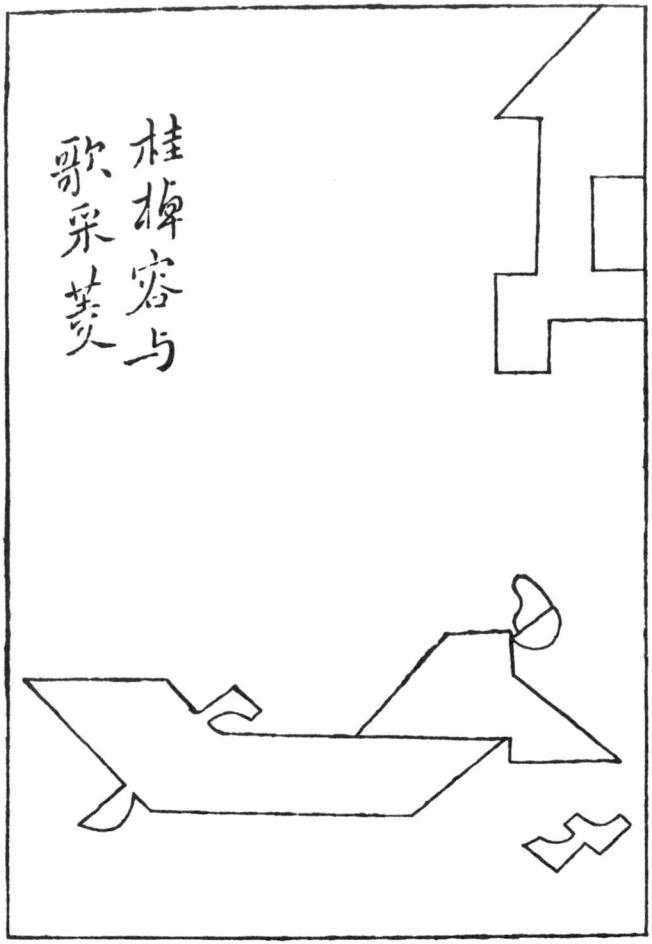

330

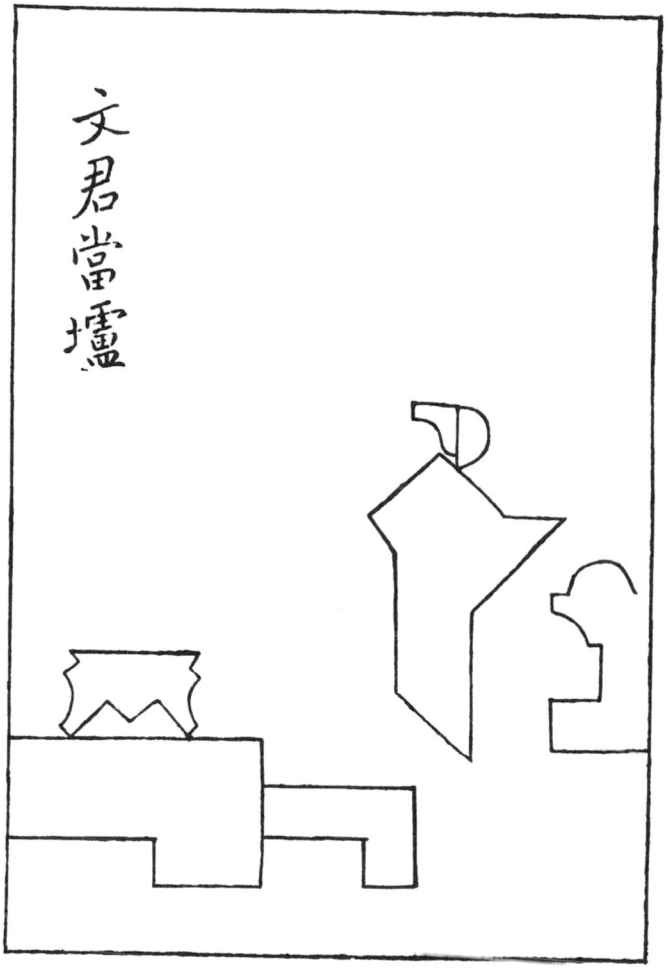

331

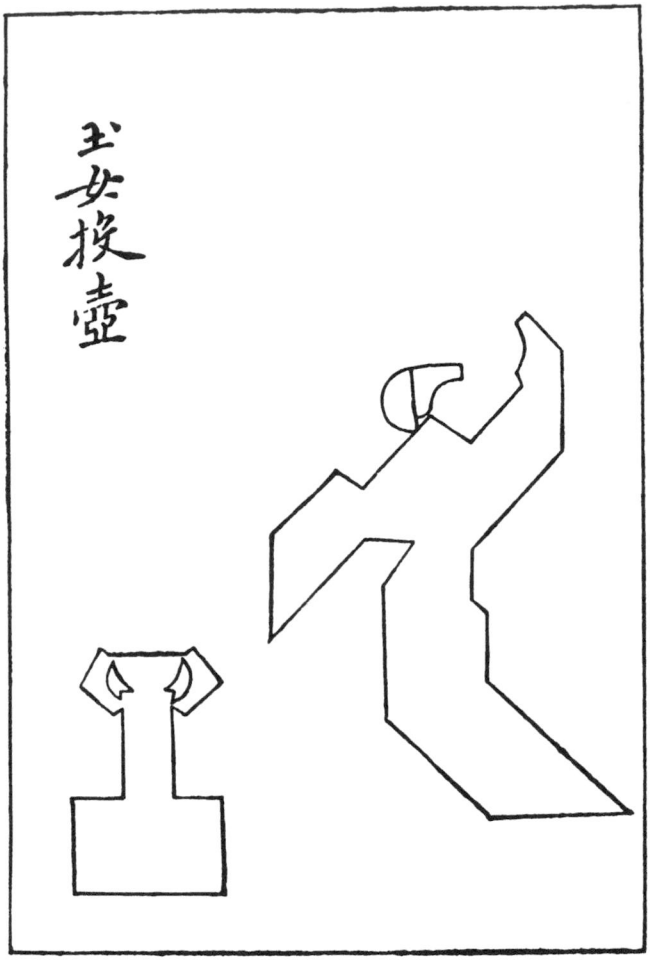

332

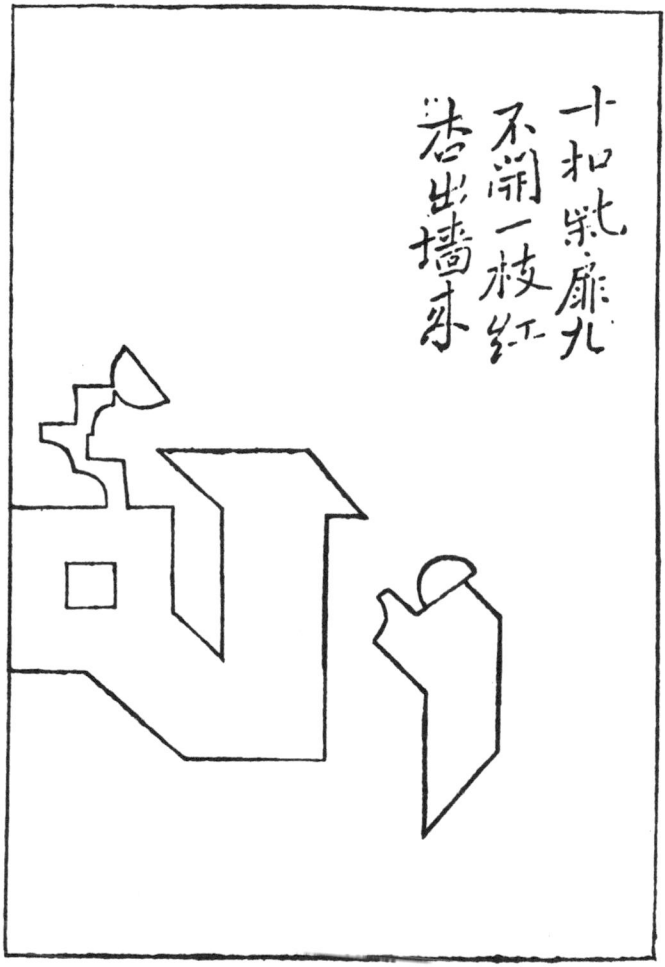
一扣就扉九
不開一枝紅
杏出墻來

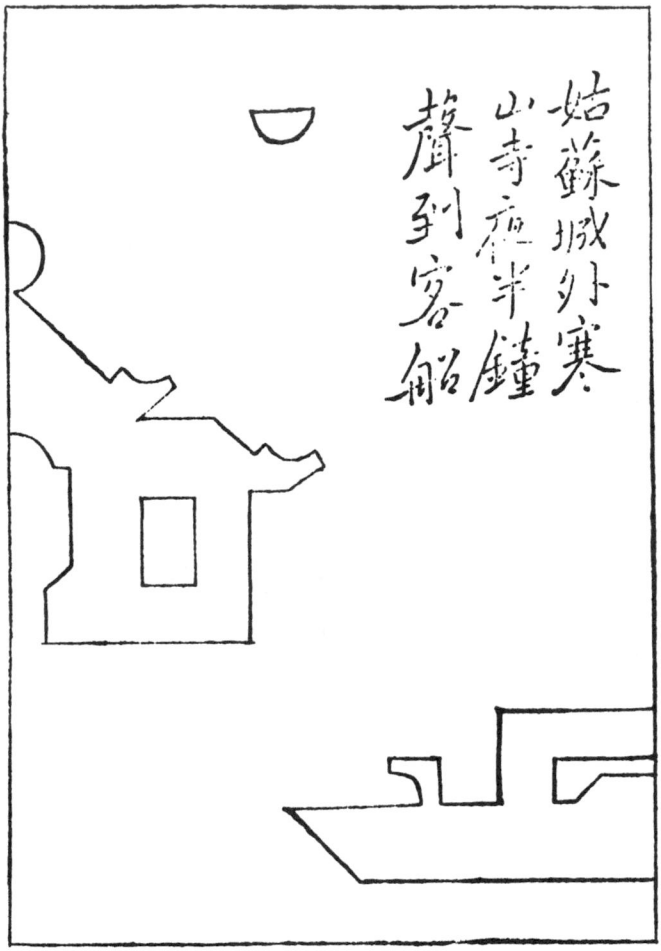

334

Solutions

1

2

3

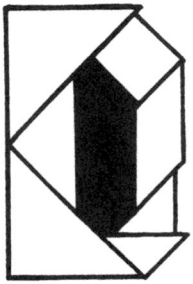

4

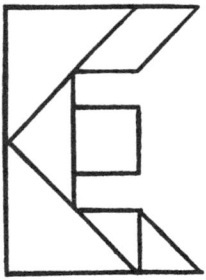

5

6

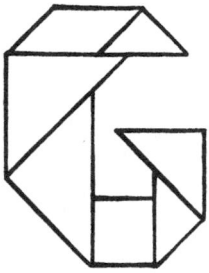

7

8

9

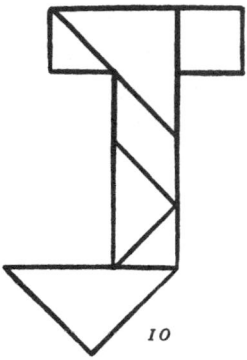

10

11

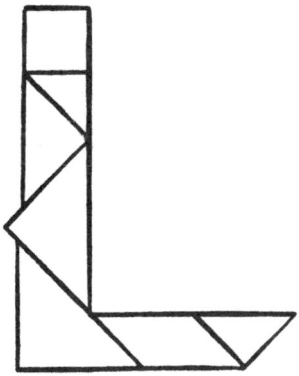

12

13

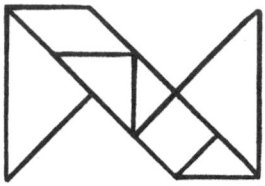

14

15

16

17

18

19

20

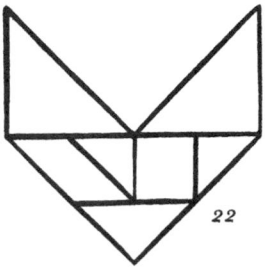

22

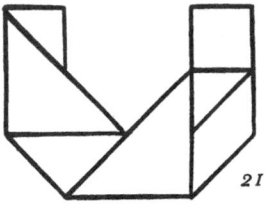

21

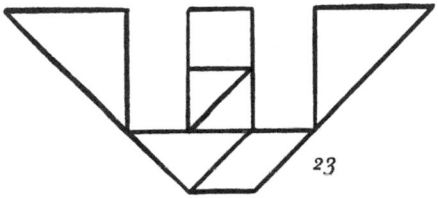

23

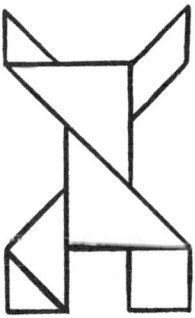

24

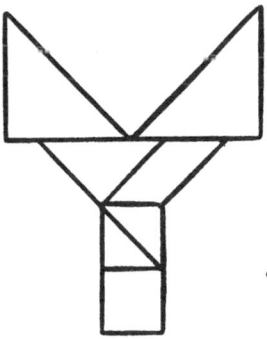

25

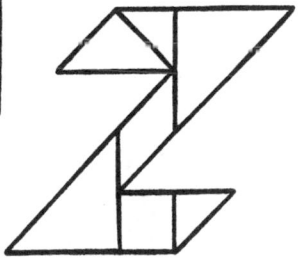

26

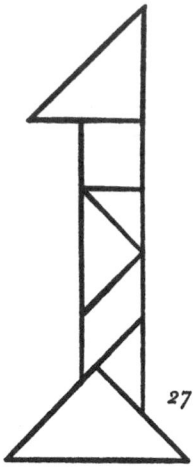

27

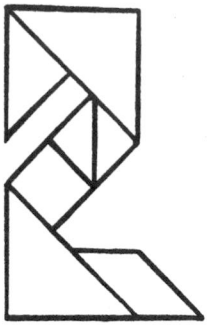

28

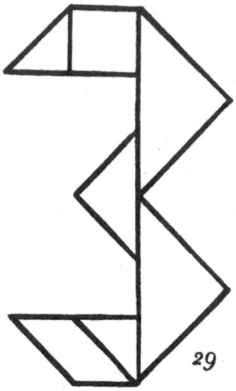

29

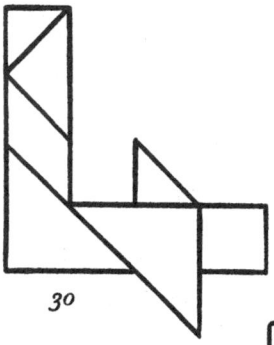

30

31

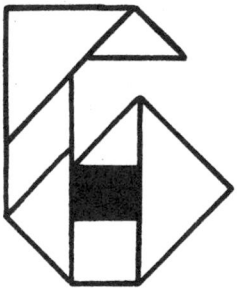

32

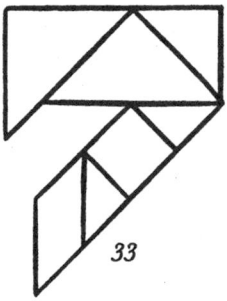

33

34

THE TANGRAM ZOO

35

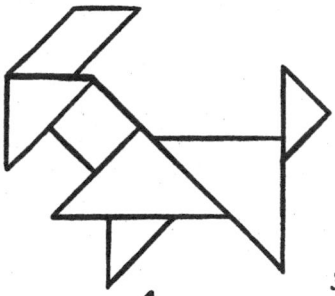

36

GOATS

Polar Bear

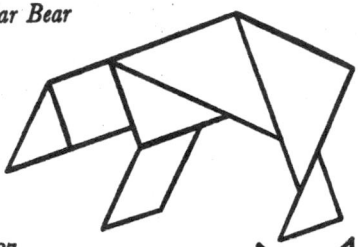

37

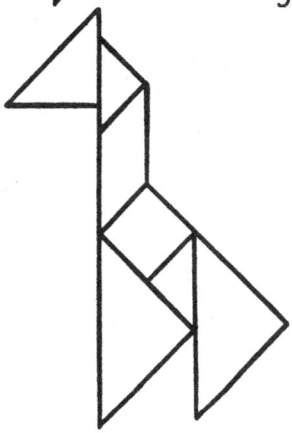

38 *Giraffe*

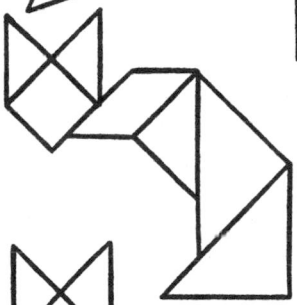

40

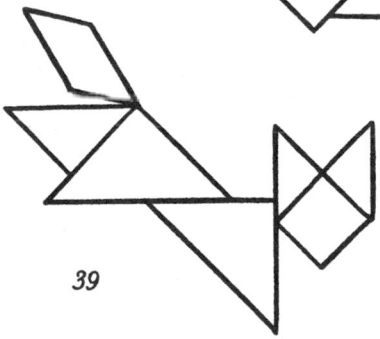

39

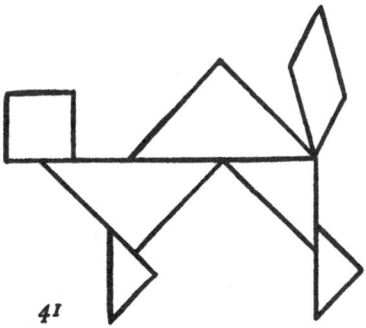

41

CATS

CATS AND DOGS

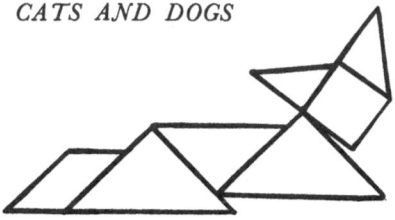

42

43

44

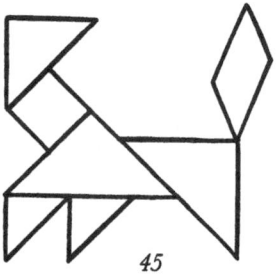

45

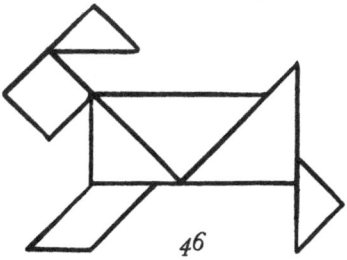

46

47 Camel

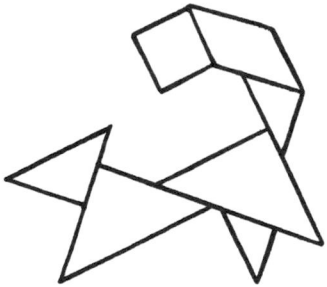

48 Squirrel

BIRDS

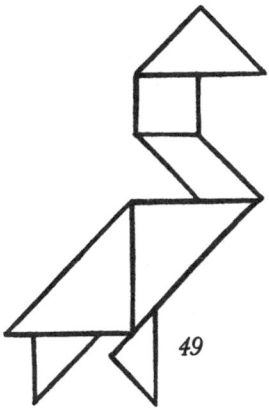

49

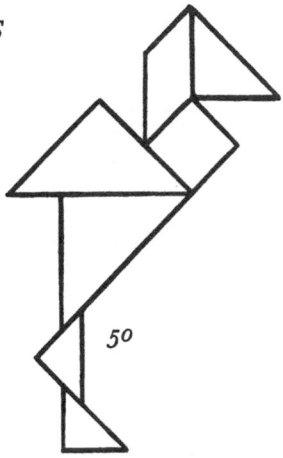

50

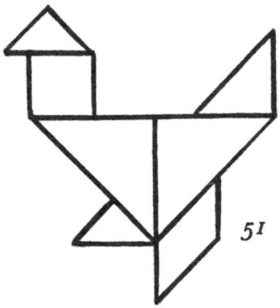

51

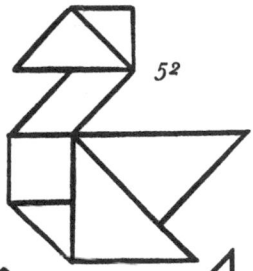

52

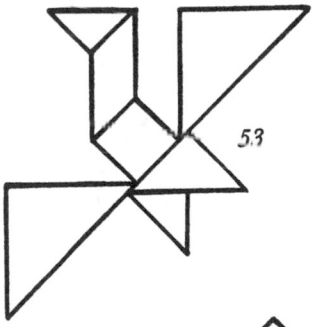

53

54

55

MORE BIRDS

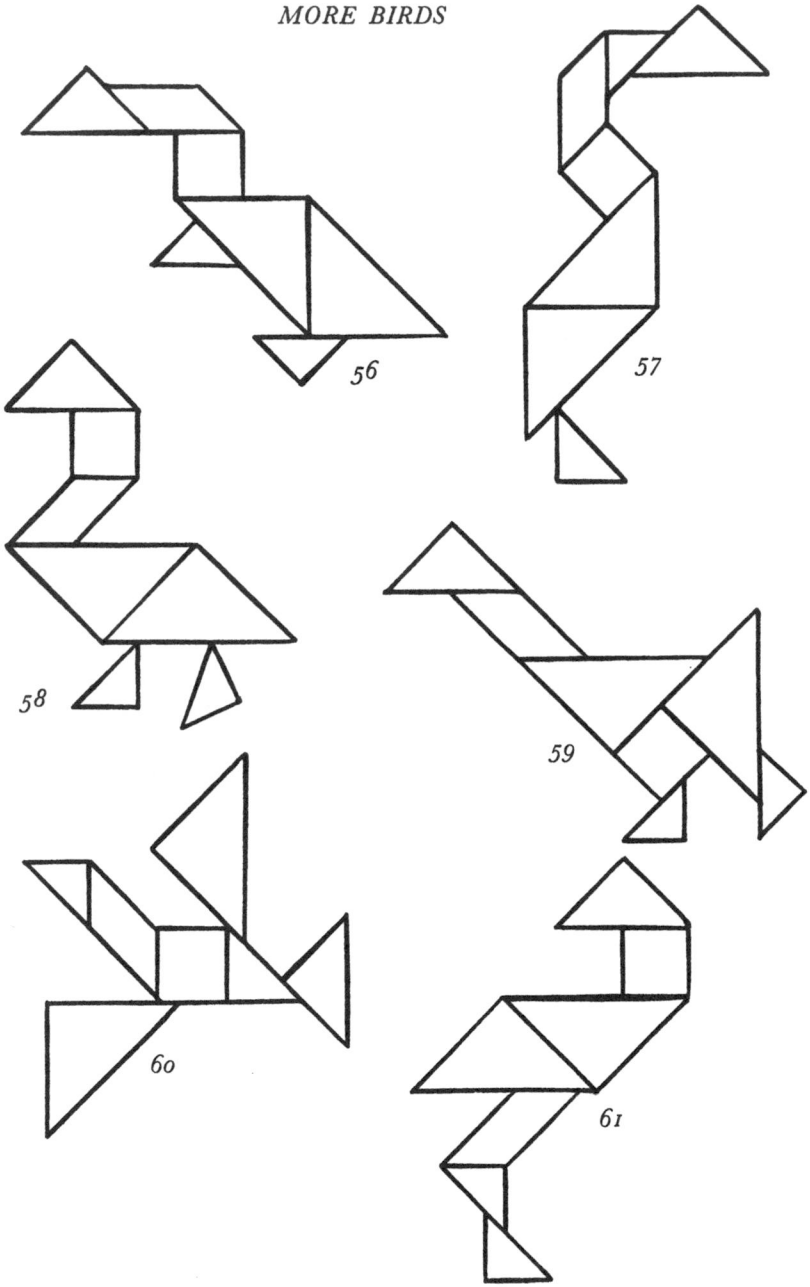

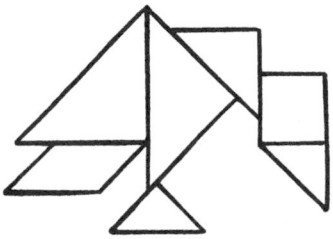

62

63

VULTURES

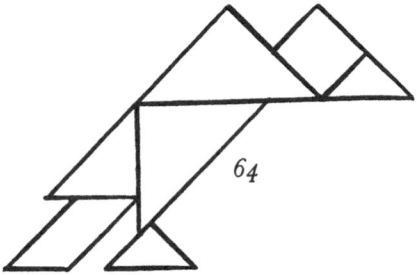

64

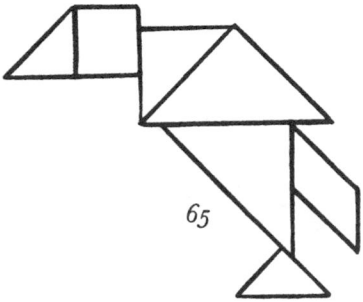

65

66

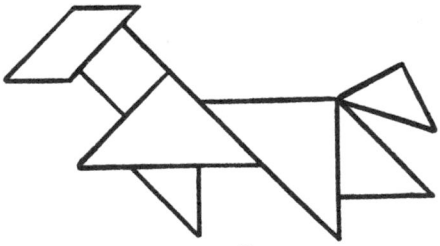

67

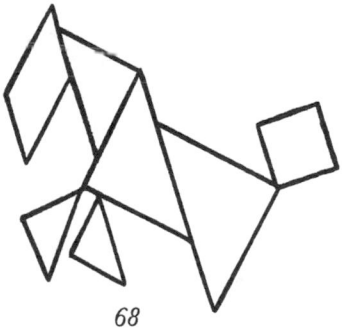

68

HORSES

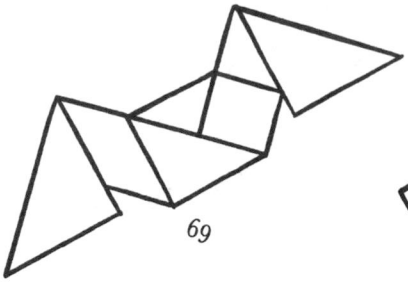

69

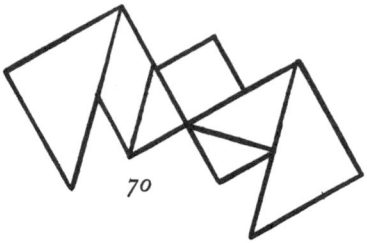

70

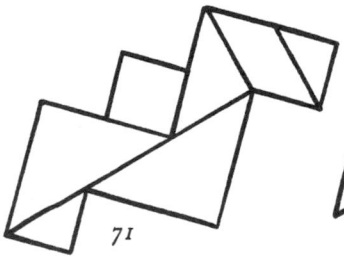

71

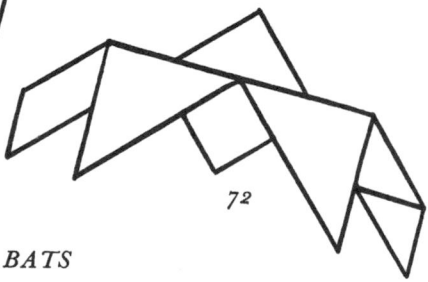

72

BATS

73 *Kangaroo*

74 *Horse*

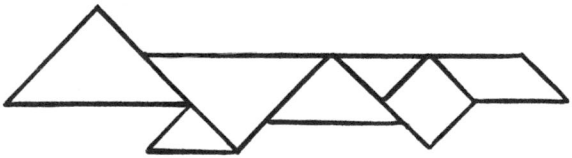

75 *Crocodile*

SEA CREATURES

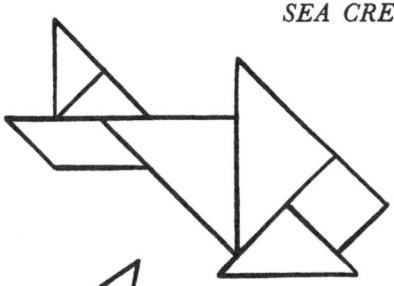

76

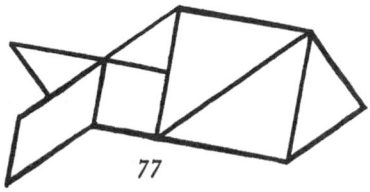

77

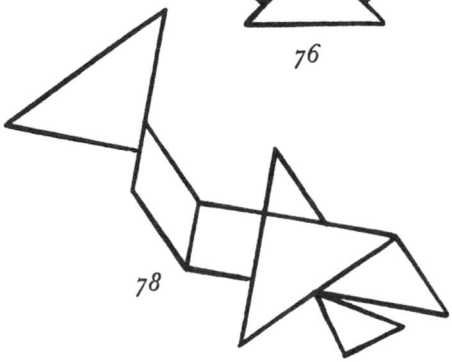

78

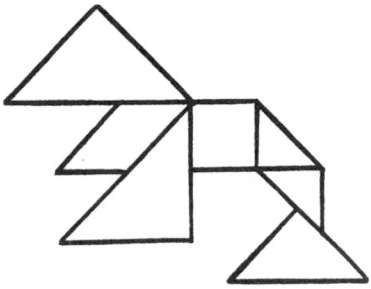

79 Lobster

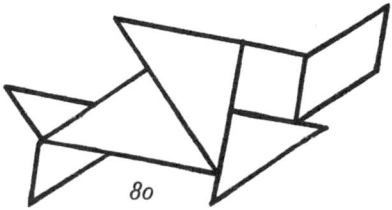

80

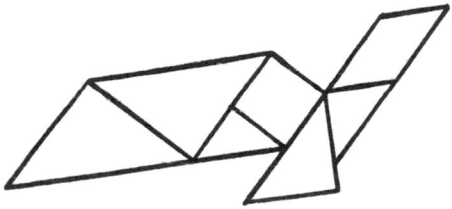

81 Turtle

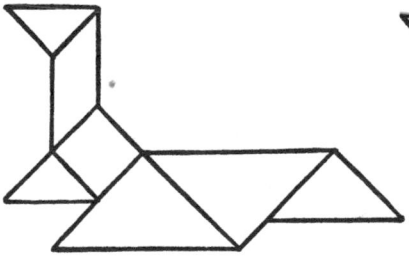

82 Seal

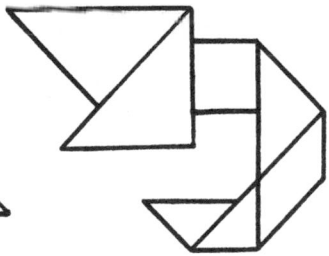

83 Shrimp

SEA CREATURES

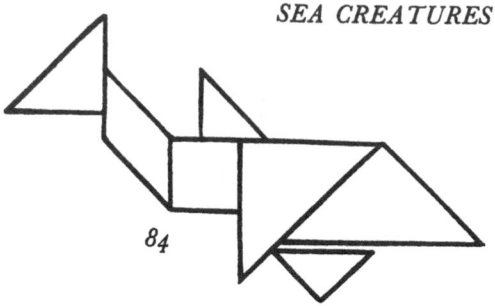

84

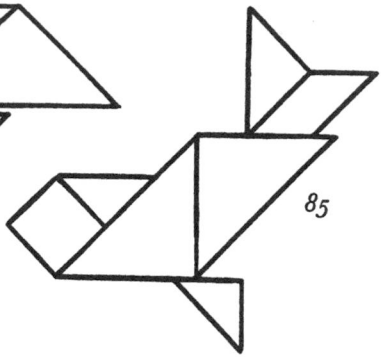

85

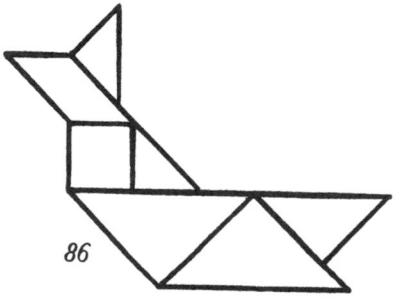

86

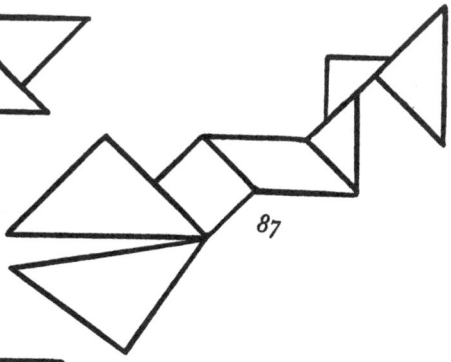

87

88

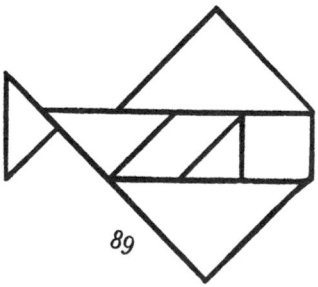

89

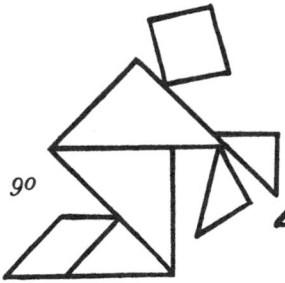

90

91

Shoeshine Boy and Customer

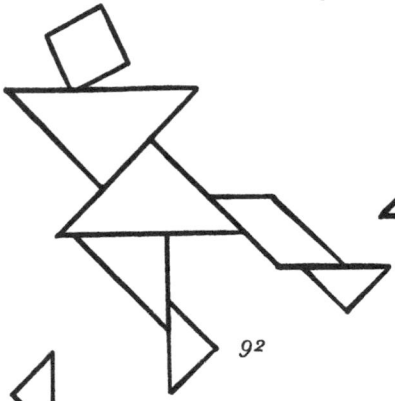

92

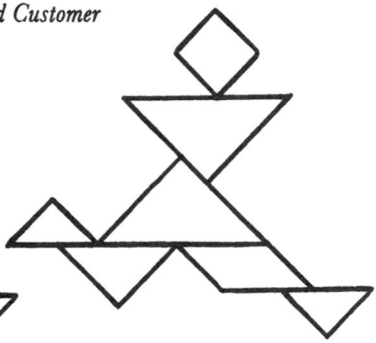

93

94

ACROBATS

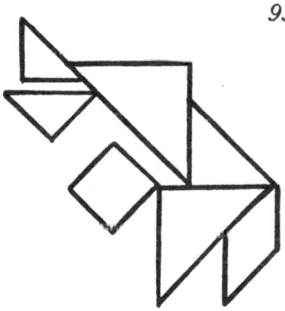

95

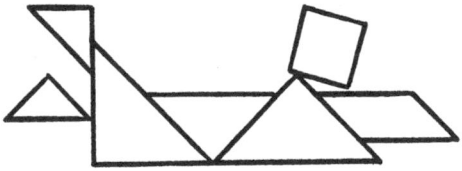

96

97

98

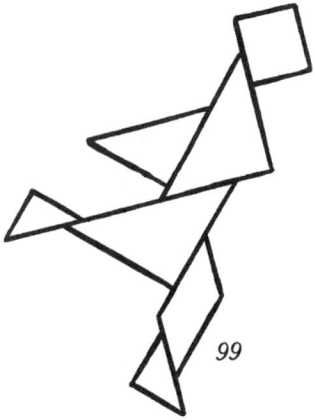

99

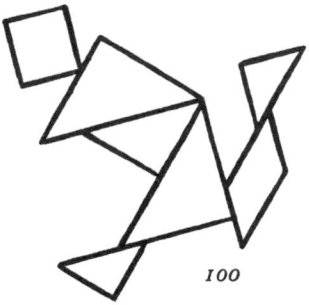

100

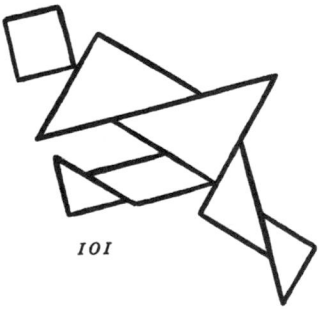

101

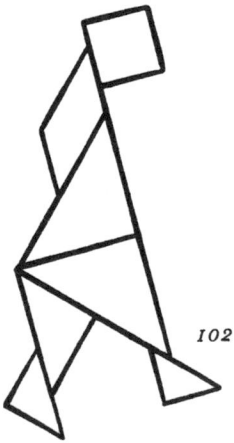

102

MEDIEVAL LADIES

103

104

105

106

AND SERVANTS

107

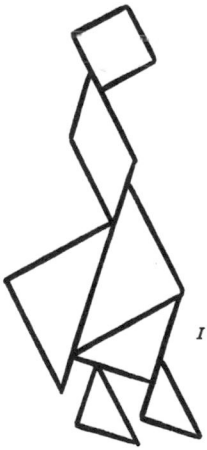

108

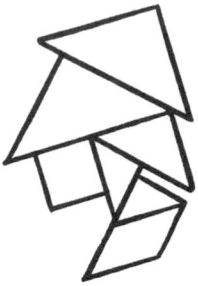

109

110

111

112

113

114

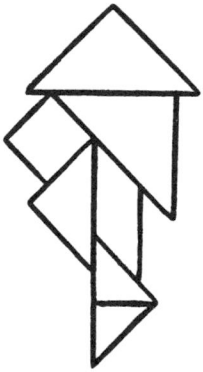

115

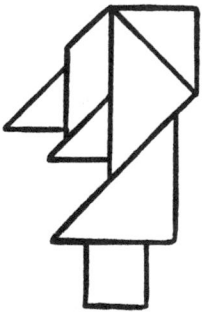

116

117

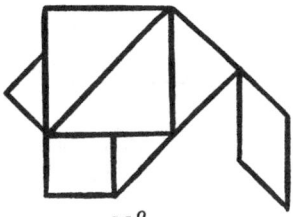

118

119

120

121

122

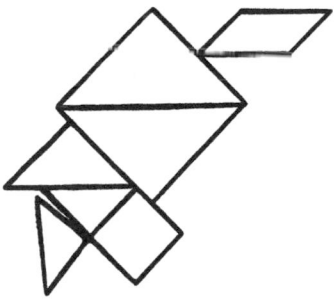

123

124

125

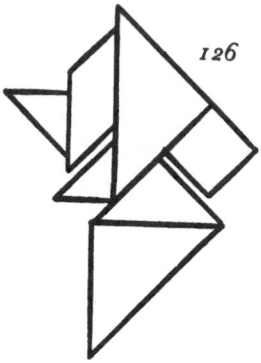

126

127

128

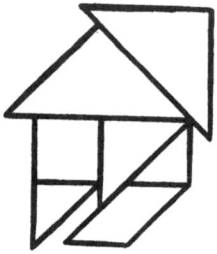

129

130

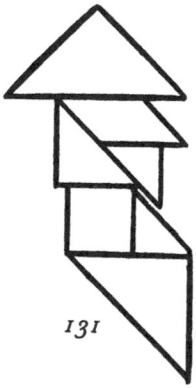

131

132

133

134

135

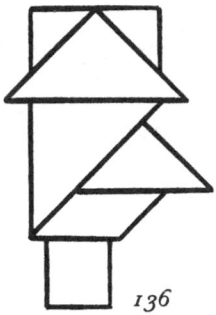

136

137

138

HORSEMEN

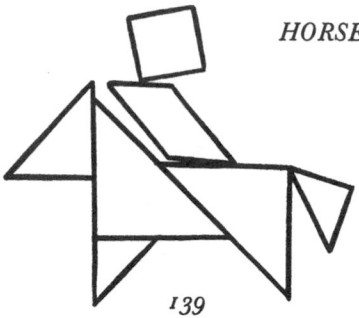

139

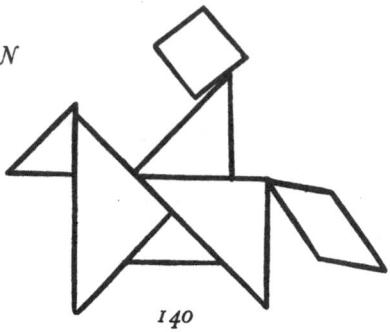

140

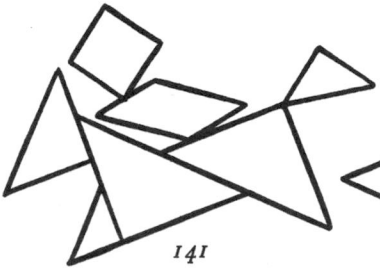

141

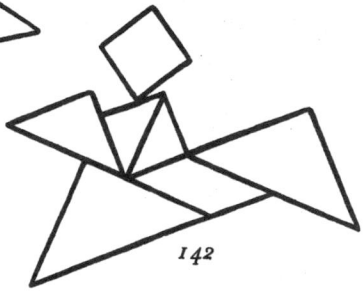

142

143

144

145

146 Pipe

147

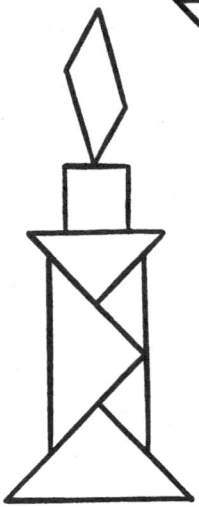

148 Candle

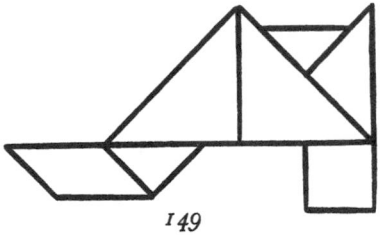

149

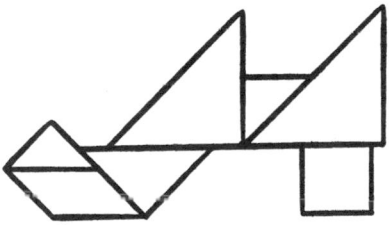

150

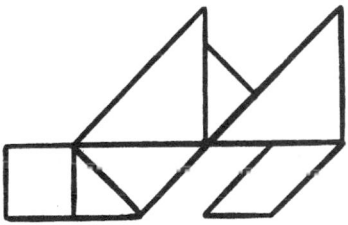

151

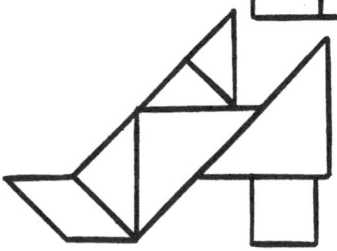

152

SHOES

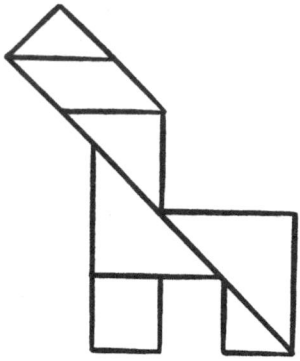

153

154

CHAIRS

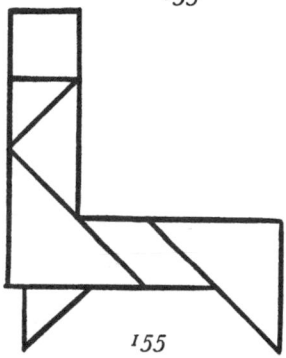

155

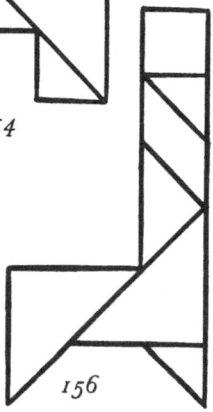

156

157

158 *Shirt*

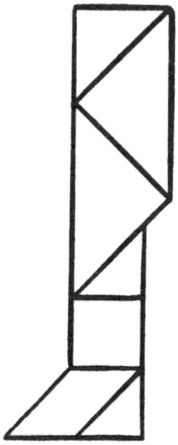

159 *Stocking*

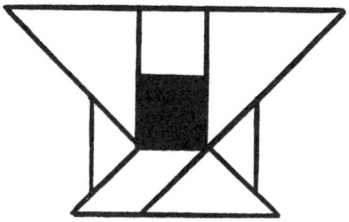

160

161 *Watering can*

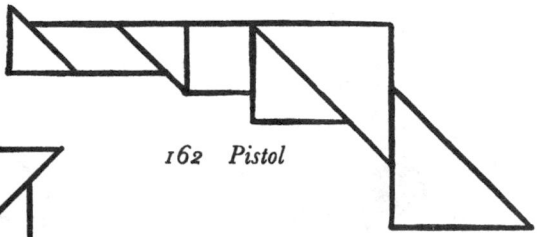

162 *Pistol*

164

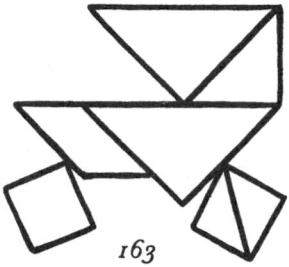

163

BABY CARRIAGES

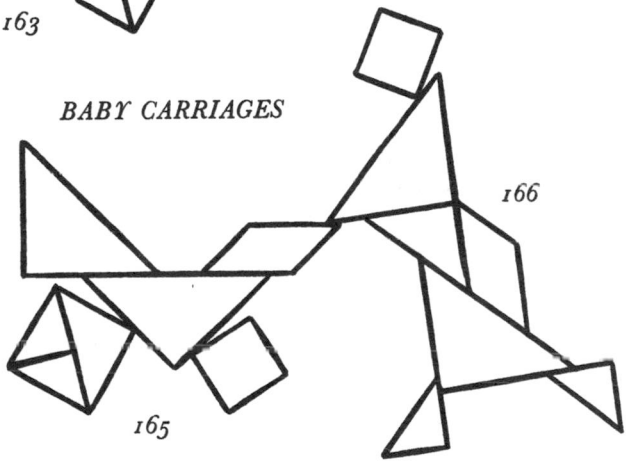

166

165

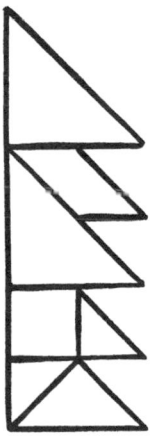

167 ???

BOATS

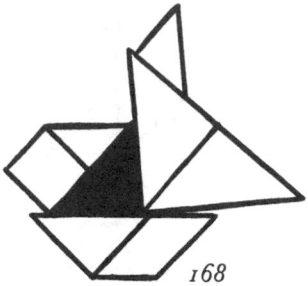

168

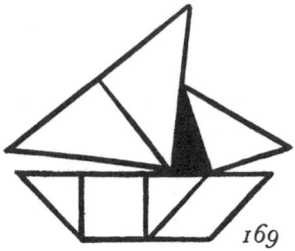

169

170

171

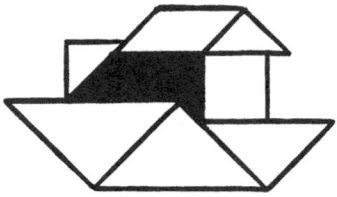

172

173

AND BRIDGES

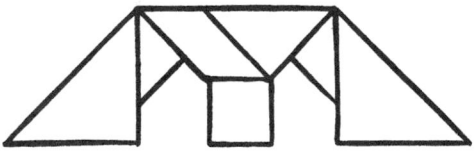

174

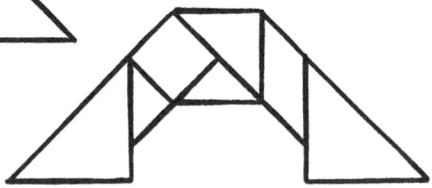

175

BRIDGES

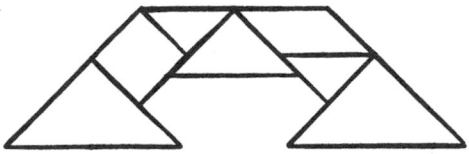

176

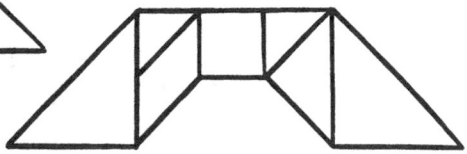

177

178

179

180

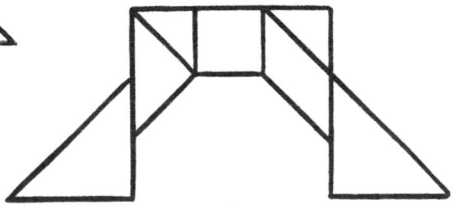

181

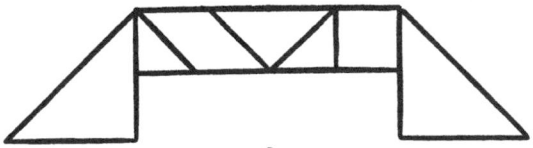

182

BOATS

183

184

185

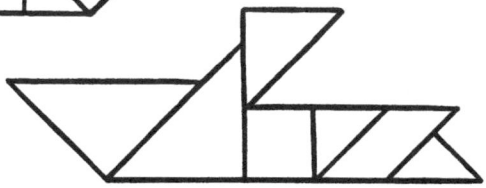

186

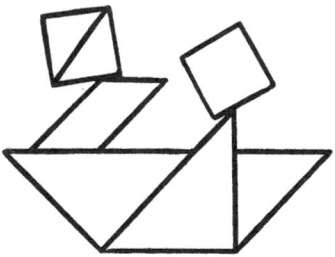

187 & 188 Rowing to the lighthouse

THE HOUSE THAT JACK BUILT

Illustrated by Sam Loyd

189

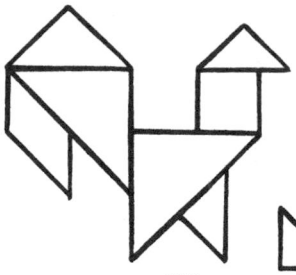

190

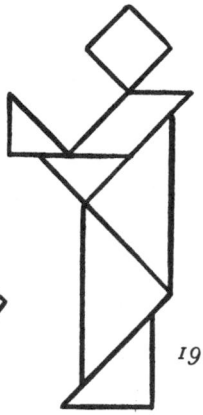

191

192

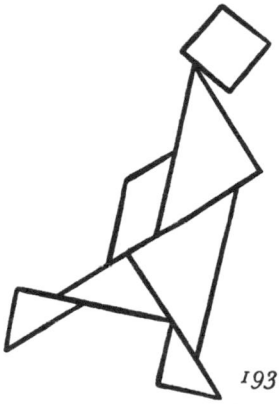

193

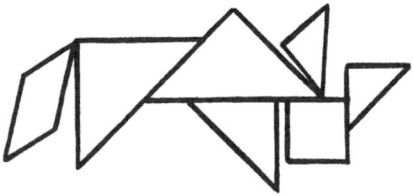

194

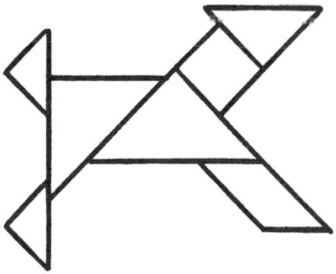

195

196

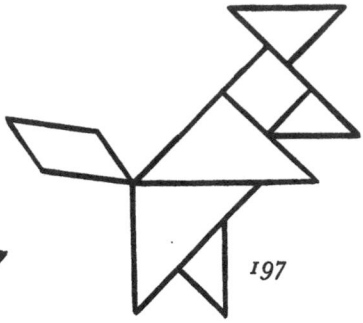

197

198

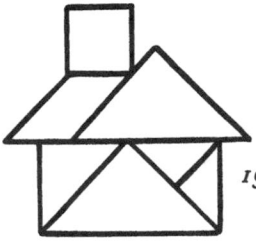

199

201

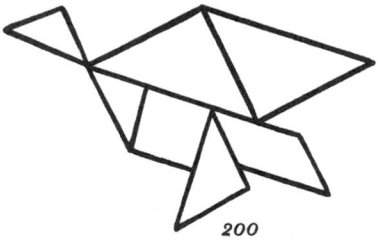

200

202

THE STORY OF CINDERELLA

Illustrated by Sam Loyd

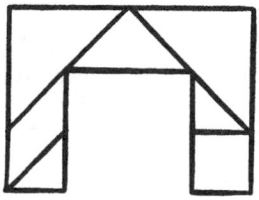

203

204

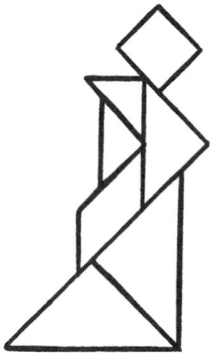

205

206

207

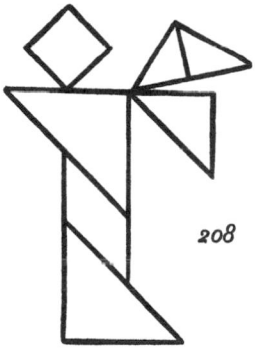

208

209

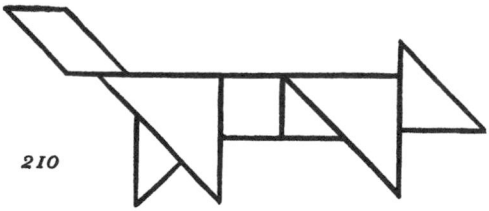

210

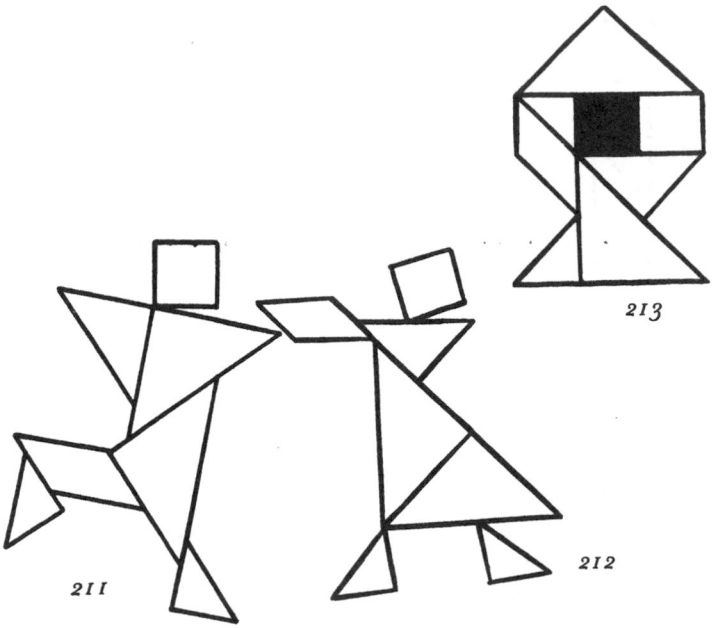

211 212 213

214

215

216

A GAME OF BILLIARDS

By H. E. Dudeney

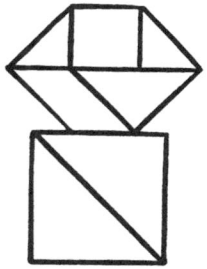

217

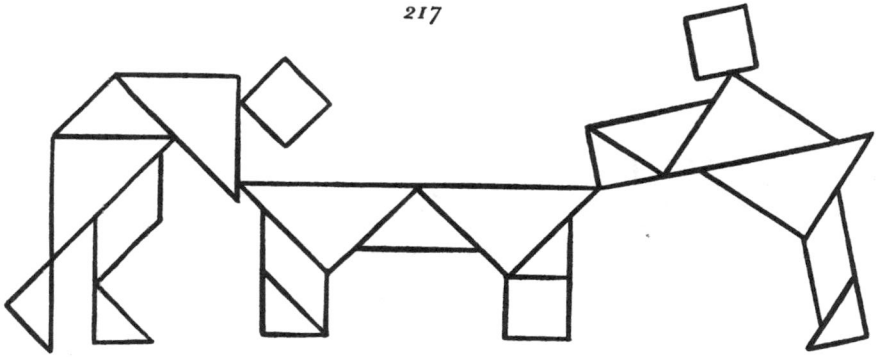

218 219 220

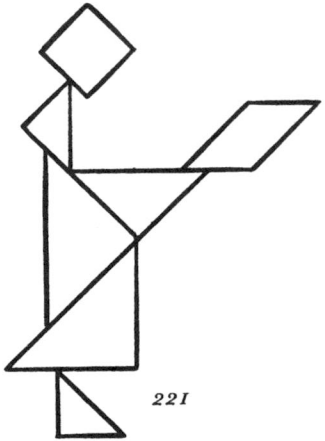

221

222

223

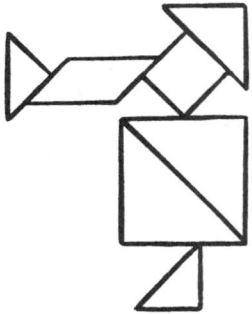

225

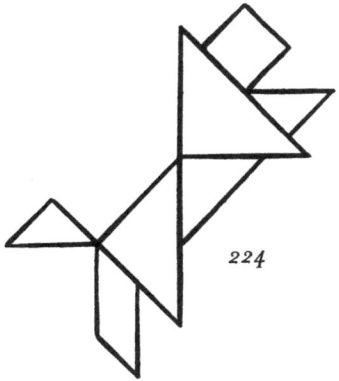

224

THE ORCHESTRA

By H. E. Dudeney

226

227

228

229

230

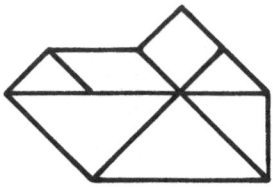

231

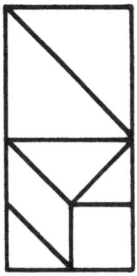

232

233

234

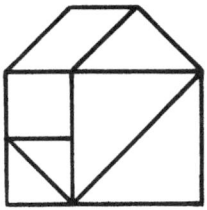

235

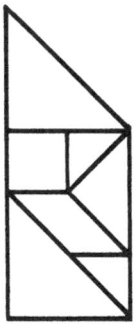

236

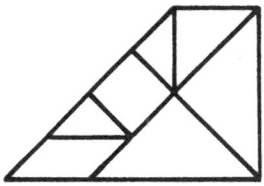

237

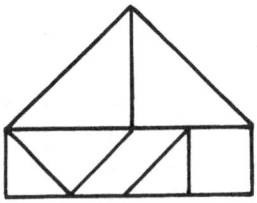

238

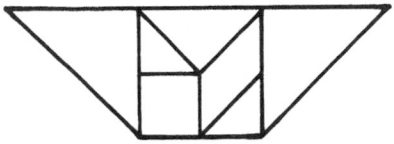

239

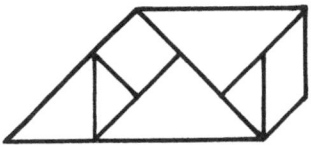

240

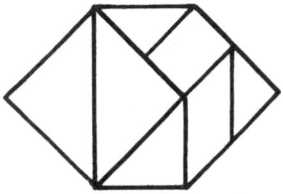

241

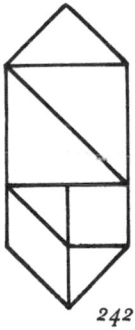

242

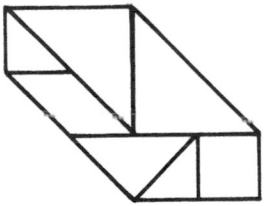

243

SOME CHINESE CHARACTERS

244

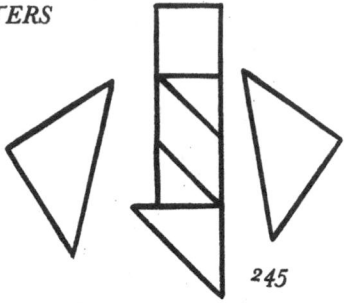

245

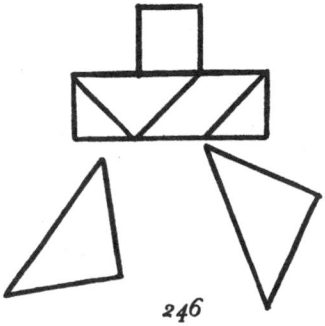

246

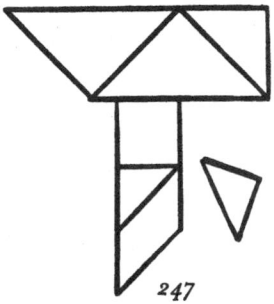

247

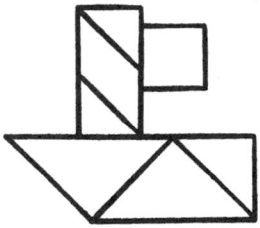

248

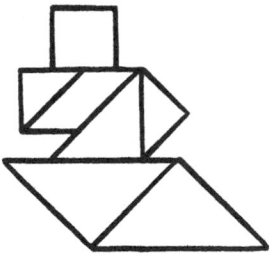

249

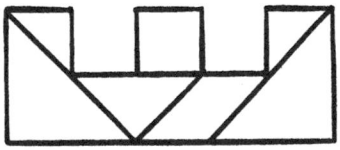

250

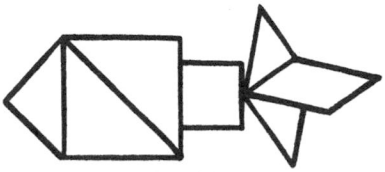

251 Goldfish

252 Riding a crane

253 Bottle gourd

254 "Buddha's hand"

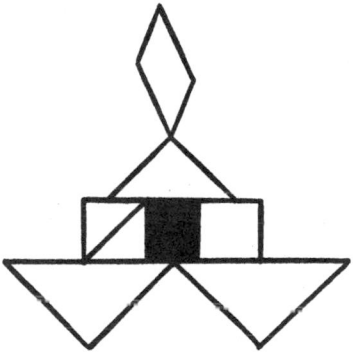

255 Helmet

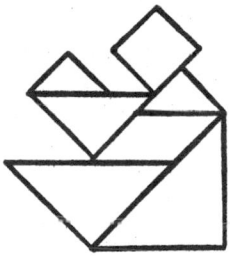

256 "Drinking alone"

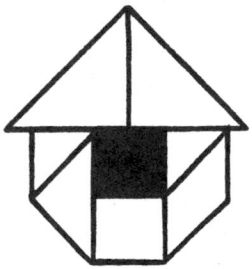

257 Clock

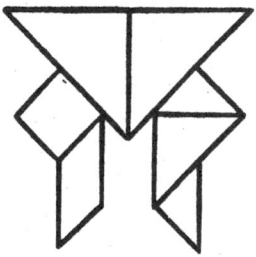

258 Brazier

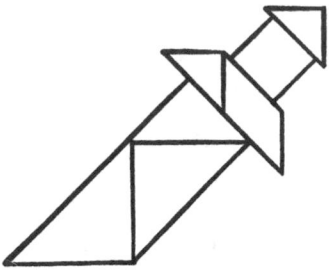

259 Dagger

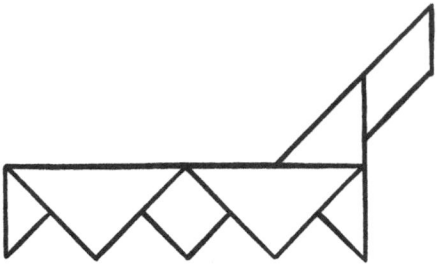

260 Pinking knife

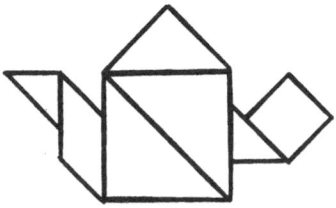

261 Teapot

262 Mulberry hook

263 Raincoat

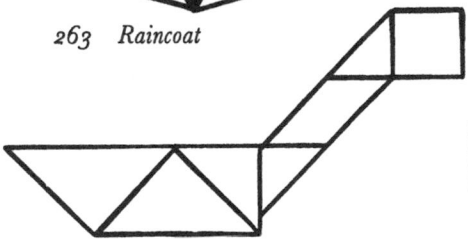

264 Soup spoon

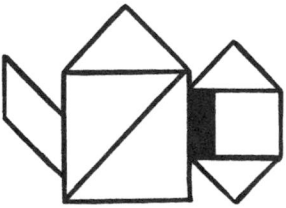

265 Teapot

266 Dragon boat

267 Three-legged toad

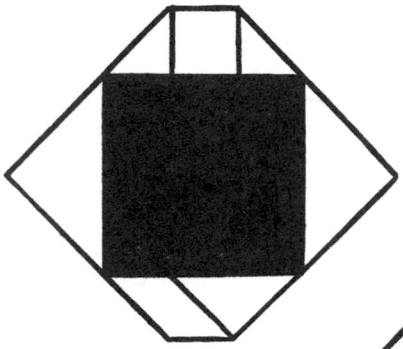

268 Pendant

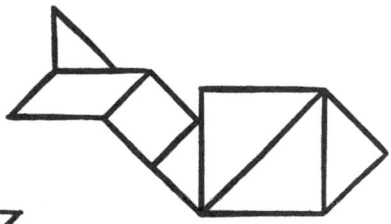

269 Carp

270 Flower bowl

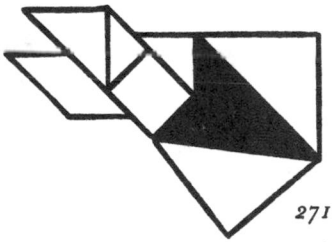

271 Scissors

272 Saw

273 Clothes post

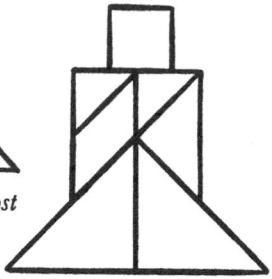

274 Bell

275 Feather tube

276 Lucky charm

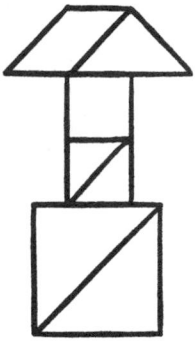

277 Look-out tower

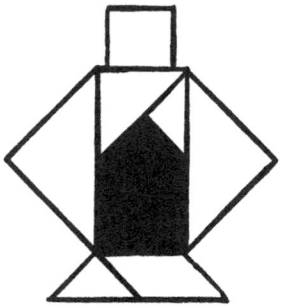

278 Medicine jar

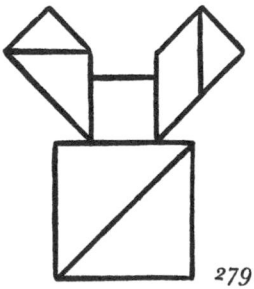

279 Mirror rack

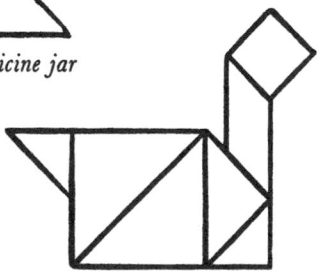

280 Ladle

TWO BASKETS

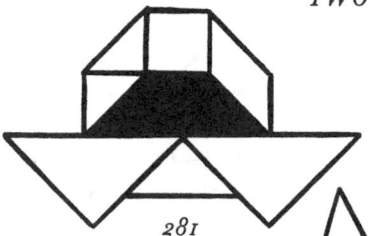

281

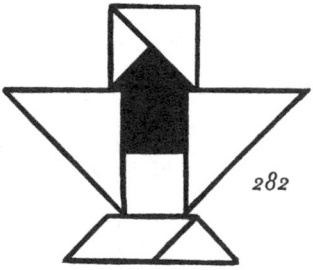

282

283
Arrow

284 Candle

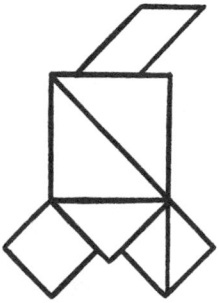

285 Water pot

287 Mallet

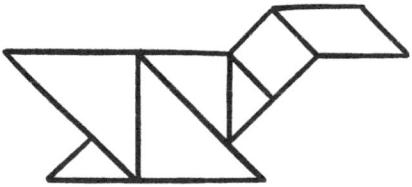

286 Measure

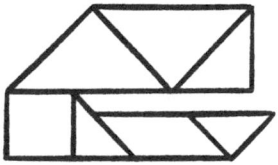

288 Penknife

For a resolution of the paradox in Tangrams 289 and 290 I cannot do better than repeat Dudeney's explanation: "It will be noticed that in both cases the head, hat and arm are precisely alike, and the width at the base of the body is the same. But this body contains four pieces in the first case, and in the second design only three. The first is larger than the second by exactly that narrow strip indicated by the dotted line between A and B. This strip is therefore exactly equal in area to the piece forming the foot in the other design, though when thus distributed along the side of the body the increased dimension is not easily apparent to the eye."

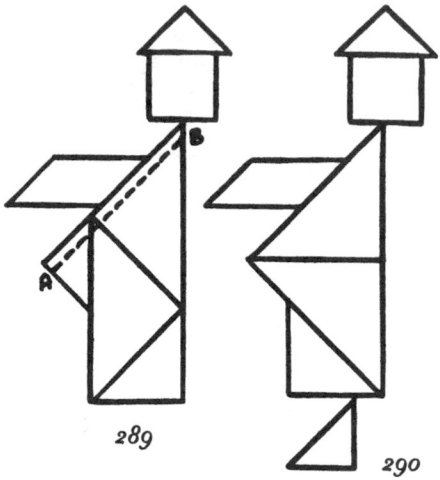

289 *290*

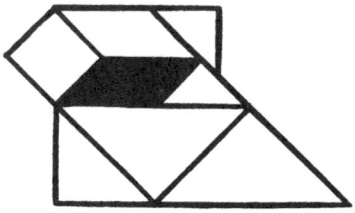

291

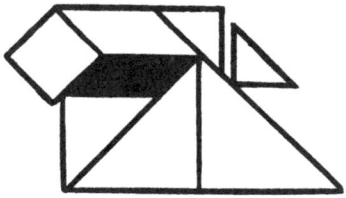

292

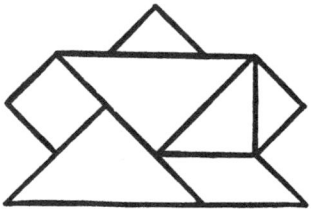

293

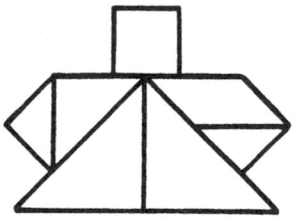

294

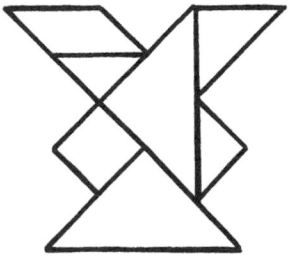

295

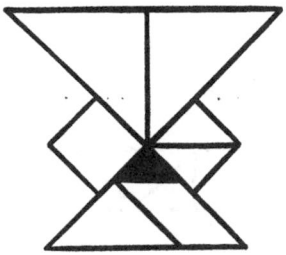

296

297

298

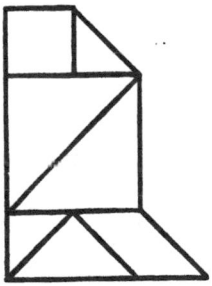

299

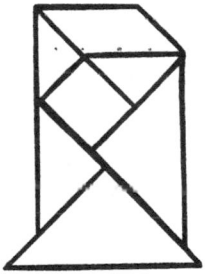

300

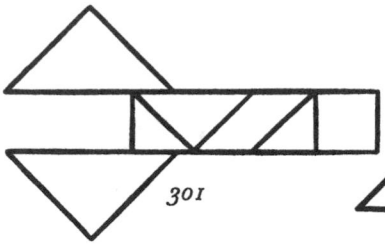

301

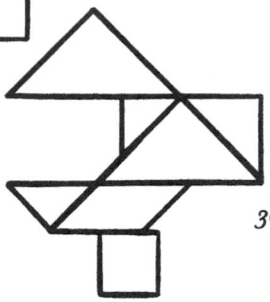

302

303

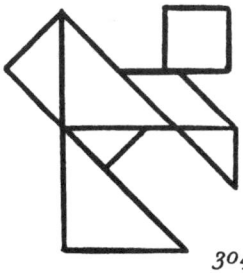

304

305

306

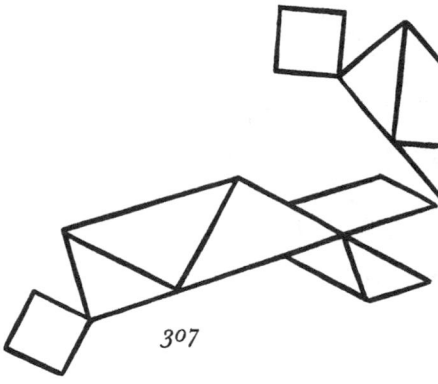

307

308

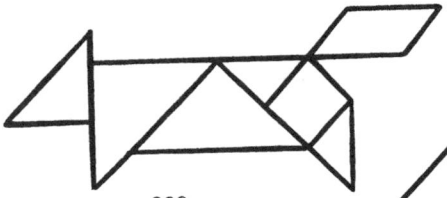

309

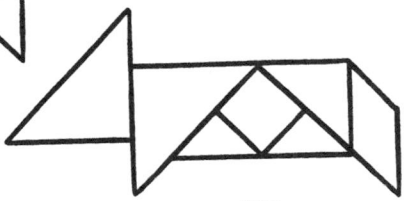

310

311

312 House

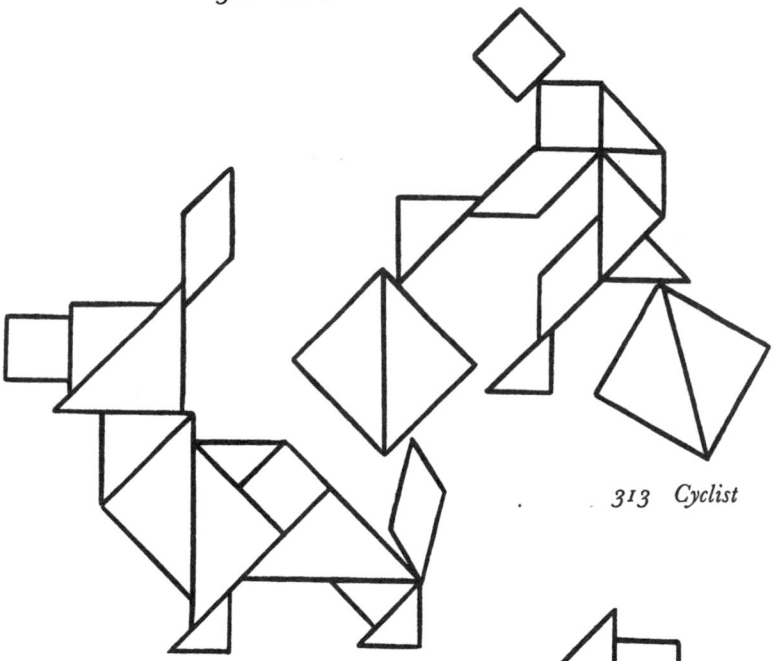

313 Cyclist

314 Dog

315 Lion

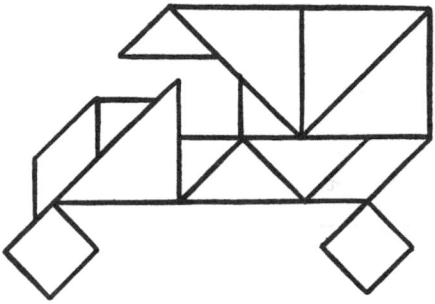

316 Car (ancient vintage)

317 Sports car

318 Coach

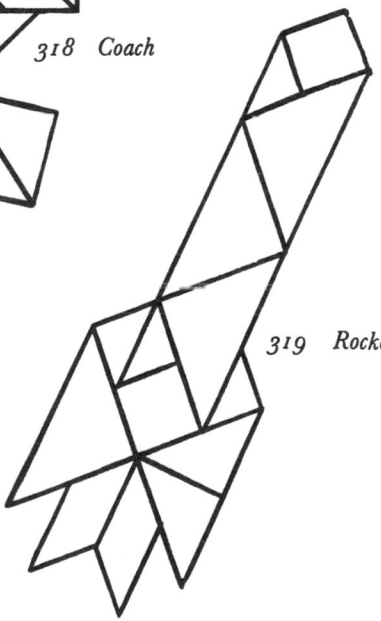

319 Rocket

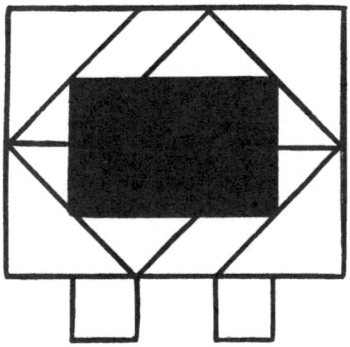

320 Television set

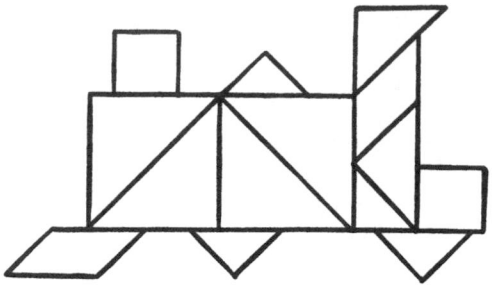

321 Locomotive

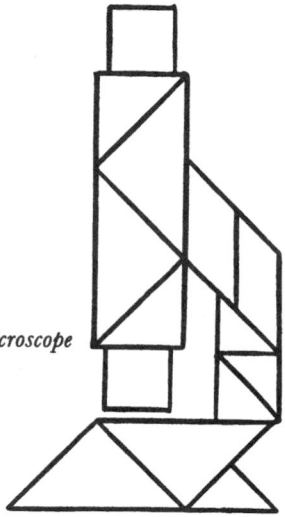

322 Microscope

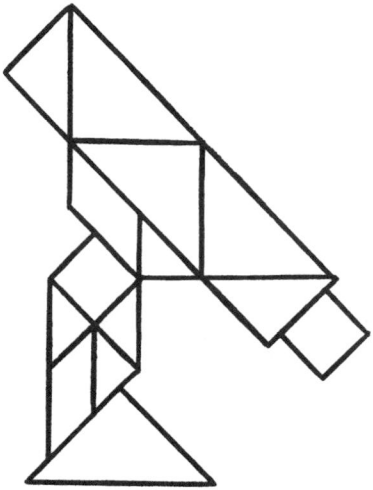

323 Telescope

324 Alarm clock

325

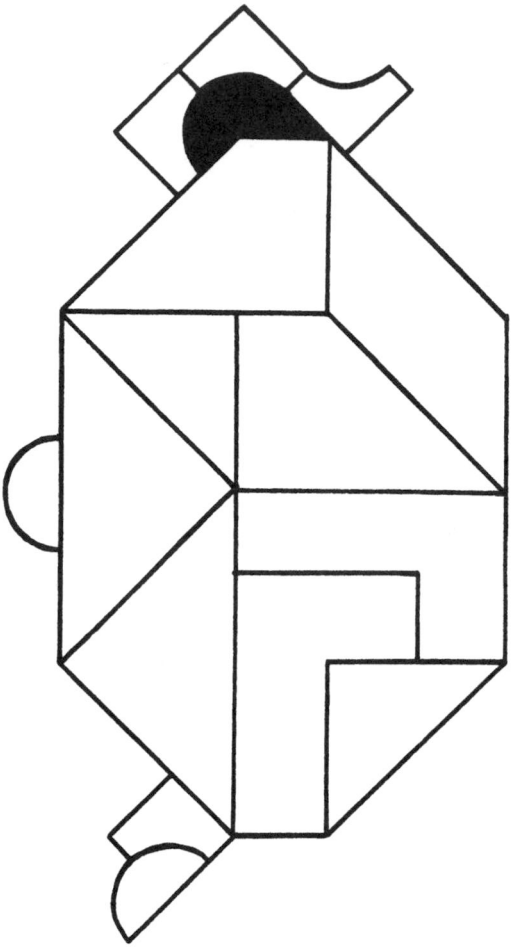

326

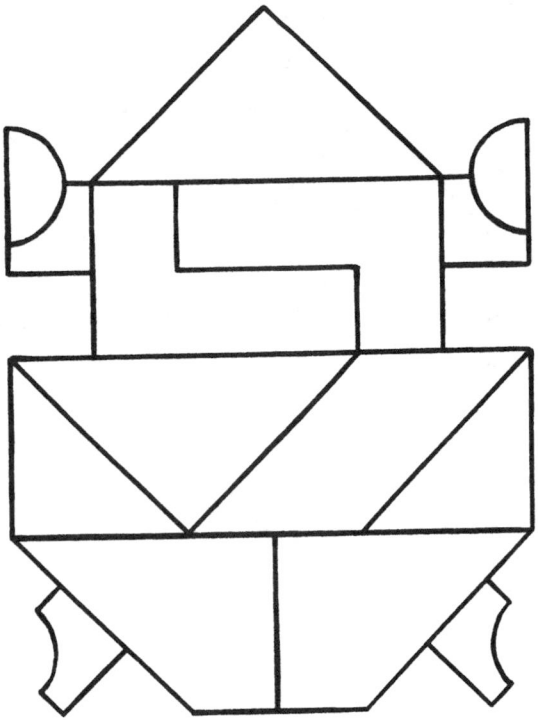

327

328

329

330

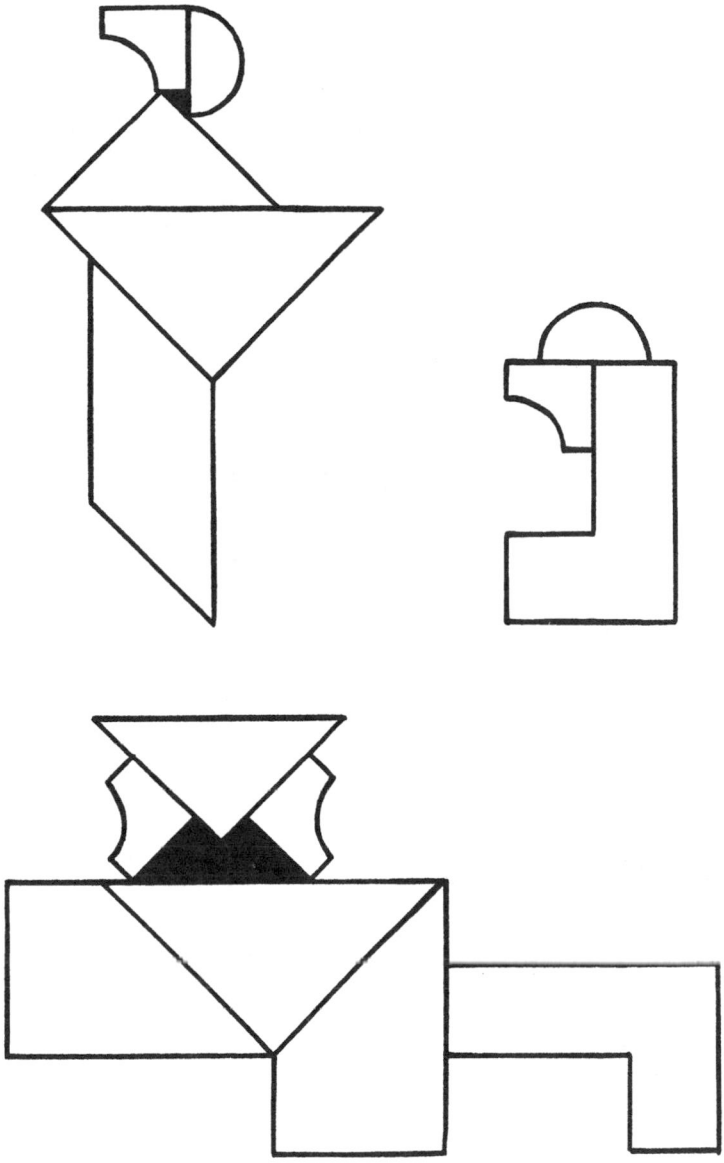

331

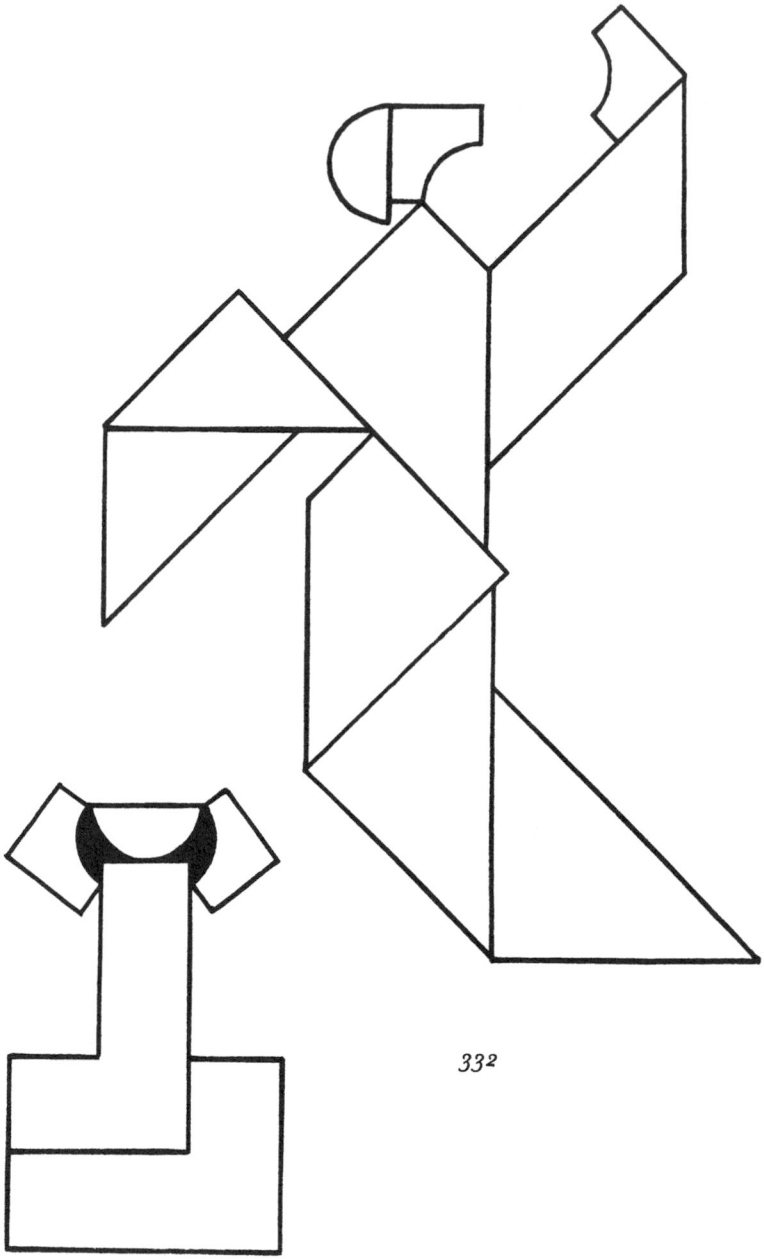

332

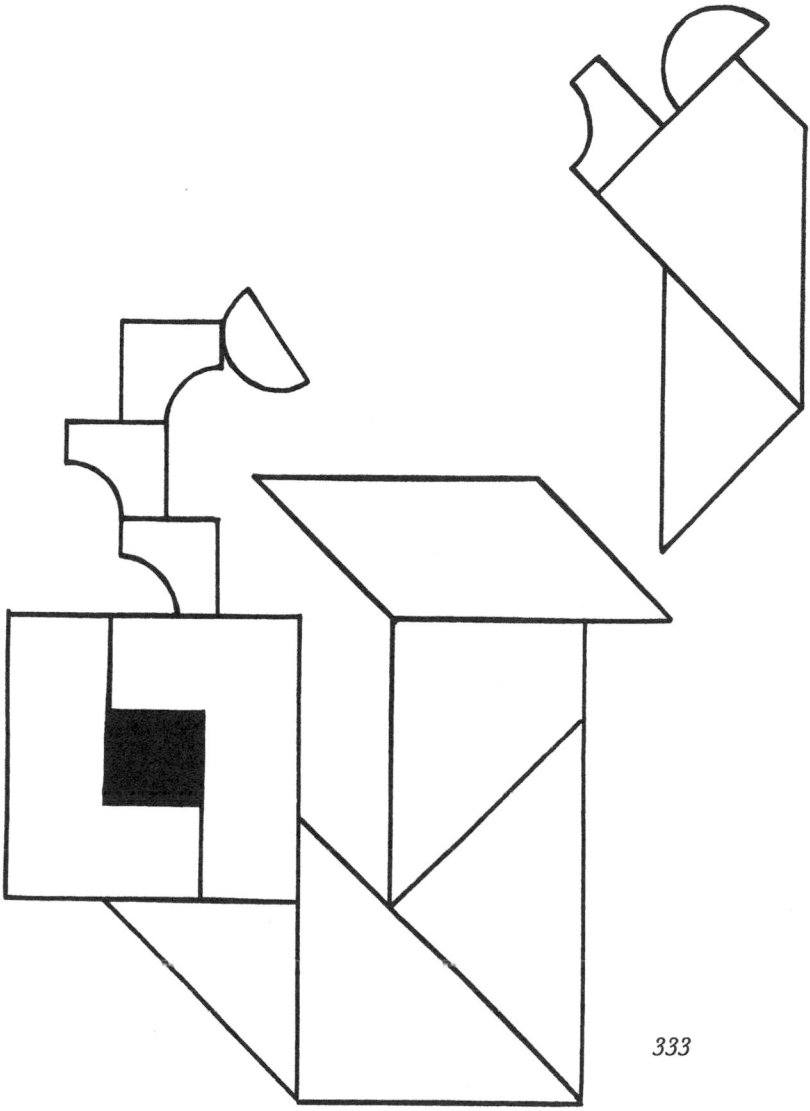

333

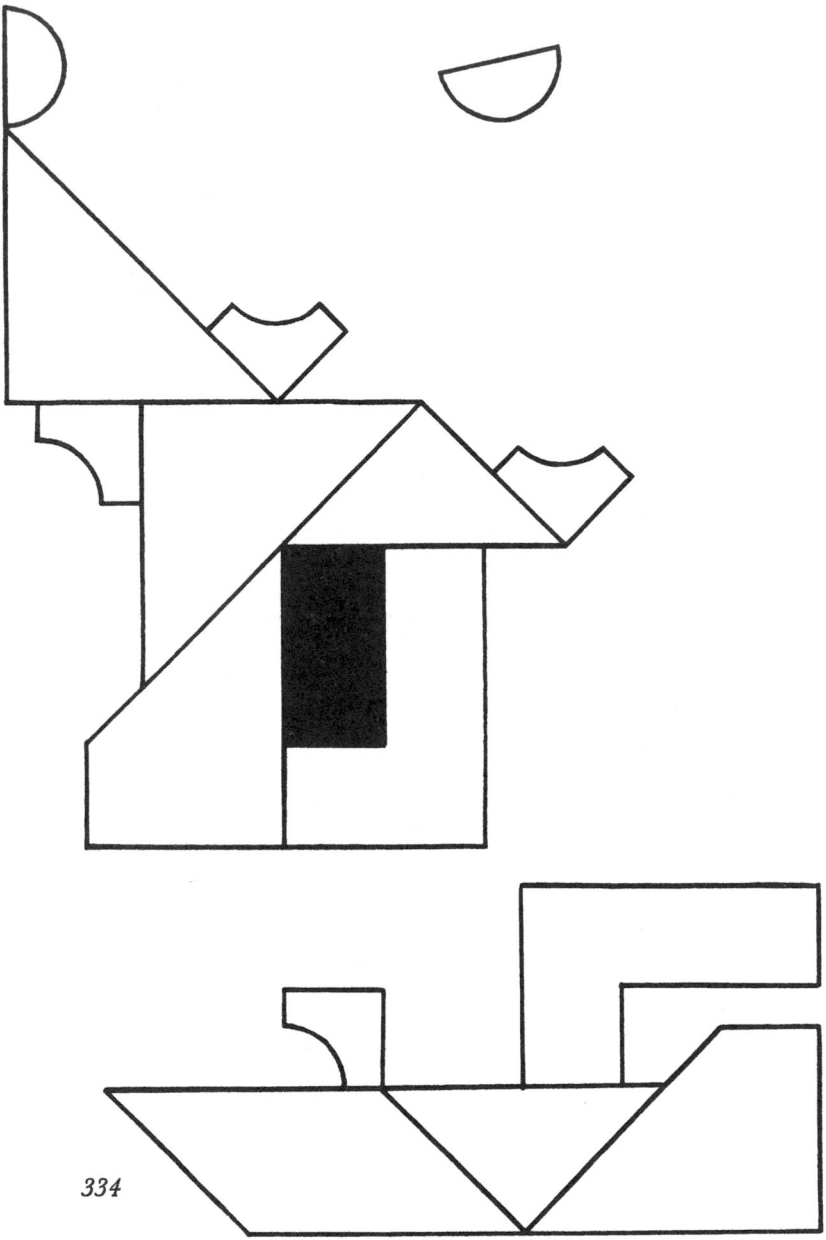

334

A CATALOG OF SELECTED

DOVER BOOKS

IN ALL FIELDS OF INTEREST

A CATALOG OF SELECTED DOVER BOOKS IN ALL FIELDS OF INTEREST

100 BEST-LOVED POEMS, Edited by Philip Smith. "The Passionate Shepherd to His Love," "Shall I compare thee to a summer's day?" "Death, be not proud," "The Raven," "The Road Not Taken," plus works by Blake, Wordsworth, Byron, Shelley, Keats, many others. 96pp. 5 3/16 x 8 1/4. 0-486-28553-7

100 SMALL HOUSES OF THE THIRTIES, Brown-Blodgett Company. Exterior photographs and floor plans for 100 charming structures. Illustrations of models accompanied by descriptions of interiors, color schemes, closet space, and other amenities. 200 illustrations. 112pp. 8 3/8 x 11. 0-486-44131-8

1000 TURN-OF-THE-CENTURY HOUSES: With Illustrations and Floor Plans, Herbert C. Chivers. Reproduced from a rare edition, this showcase of homes ranges from cottages and bungalows to sprawling mansions. Each house is meticulously illustrated and accompanied by complete floor plans. 256pp. 9 3/8 x 12 1/4. 0-486-45596-3

101 GREAT AMERICAN POEMS, Edited by The American Poetry & Literacy Project. Rich treasury of verse from the 19th and 20th centuries includes works by Edgar Allan Poe, Robert Frost, Walt Whitman, Langston Hughes, Emily Dickinson, T. S. Eliot, other notables. 96pp. 5 3/16 x 8 1/4. 0-486-40158-8

101 GREAT SAMURAI PRINTS, Utagawa Kuniyoshi. Kuniyoshi was a master of the warrior woodblock print — and these 18th-century illustrations represent the pinnacle of his craft. Full-color portraits of renowned Japanese samurais pulse with movement, passion, and remarkably fine detail. 112pp. 8 3/8 x 11. 0-486-46523-3

ABC OF BALLET, Janet Grosser. Clearly worded, abundantly illustrated little guide defines basic ballet-related terms: arabesque, battement, pas de chat, relevé, sissonne, many others. Pronunciation guide included. Excellent primer. 48pp. 4 3/16 x 5 3/4. 0-486-40871-X

ACCESSORIES OF DRESS: An Illustrated Encyclopedia, Katherine Lester and Bess Viola Oerke. Illustrations of hats, veils, wigs, cravats, shawls, shoes, gloves, and other accessories enhance an engaging commentary that reveals the humor and charm of the many-sided story of accessorized apparel. 644 figures and 59 plates. 608pp. 6 1/8 x 9 1/4. 0-486-43378-1

ADVENTURES OF HUCKLEBERRY FINN, Mark Twain. Join Huck and Jim as their boyhood adventures along the Mississippi River lead them into a world of excitement, danger, and self-discovery. Humorous narrative, lyrical descriptions of the Mississippi valley, and memorable characters. 224pp. 5 3/16 x 8 1/4. 0-486-28061-6

ALICE STARMORE'S BOOK OF FAIR ISLE KNITTING, Alice Starmore. A noted designer from the region of Scotland's Fair Isle explores the history and techniques of this distinctive, stranded-color knitting style and provides copious illustrated instructions for 14 original knitwear designs. 208pp. 8 3/8 x 10 7/8. 0-486-47218-3

Browse over 9,000 books at www.doverpublications.com

ALICE'S ADVENTURES IN WONDERLAND, Lewis Carroll. Beloved classic about a little girl lost in a topsy-turvy land and her encounters with the White Rabbit, March Hare, Mad Hatter, Cheshire Cat, and other delightfully improbable characters. 42 illustrations by Sir John Tenniel. 96pp. 5 3/16 x 8 1/4. 0-486-27543-4

AMERICA'S LIGHTHOUSES: An Illustrated History, Francis Ross Holland. Profusely illustrated fact-filled survey of American lighthouses since 1716. Over 200 stations — East, Gulf, and West coasts, Great Lakes, Hawaii, Alaska, Puerto Rico, the Virgin Islands, and the Mississippi and St. Lawrence Rivers. 240pp. 8 x 10 3/4. 0-486-25576-X

AN ENCYCLOPEDIA OF THE VIOLIN, Alberto Bachmann. Translated by Frederick H. Martens. Introduction by Eugene Ysaye. First published in 1925, this renowned reference remains unsurpassed as a source of essential information, from construction and evolution to repertoire and technique. Includes a glossary and 73 illustrations. 496pp. 6 1/8 x 9 1/4. 0-486-46618-3

ANIMALS: 1,419 Copyright-Free Illustrations of Mammals, Birds, Fish, Insects, etc., Selected by Jim Harter. Selected for its visual impact and ease of use, this outstanding collection of wood engravings presents over 1,000 species of animals in extremely lifelike poses. Includes mammals, birds, reptiles, amphibians, fish, insects, and other invertebrates. 284pp. 9 x 12. 0-486-23766-4

THE ANNALS, Tacitus. Translated by Alfred John Church and William Jackson Brodribb. This vital chronicle of Imperial Rome, written by the era's great historian, spans A.D. 14-68 and paints incisive psychological portraits of major figures, from Tiberius to Nero. 416pp. 5 3/16 x 8 1/4. 0-486-45236-0

ANTIGONE, Sophocles. Filled with passionate speeches and sensitive probing of moral and philosophical issues, this powerful and often-performed Greek drama reveals the grim fate that befalls the children of Oedipus. Footnotes. 64pp. 5 3/16 x 8 1/4. 0-486-27804-2

ART DECO DECORATIVE PATTERNS IN FULL COLOR, Christian Stoll. Reprinted from a rare 1910 portfolio, 160 sensuous and exotic images depict a breathtaking array of florals, geometrics, and abstracts — all elegant in their stark simplicity. 64pp. 8 3/8 x 11. 0-486-44862-2

THE ARTHUR RACKHAM TREASURY: 86 Full-Color Illustrations, Arthur Rackham. Selected and Edited by Jeff A. Menges. A stunning treasury of 86 full-page plates span the famed English artist's career, from *Rip Van Winkle* (1905) to masterworks such as *Undine, A Midsummer Night's Dream,* and *Wind in the Willows* (1939). 96pp. 8 3/8 x 11. 0-486-44685-9

THE AUTHENTIC GILBERT & SULLIVAN SONGBOOK, W. S. Gilbert and A. S. Sullivan. The most comprehensive collection available, this songbook includes selections from every one of Gilbert and Sullivan's light operas. Ninety-two numbers are presented uncut and unedited, and in their original keys. 410pp. 9 x 12. 0-486-23482-7

THE AWAKENING, Kate Chopin. First published in 1899, this controversial novel of a New Orleans wife's search for love outside a stifling marriage shocked readers. Today, it remains a first-rate narrative with superb characterization. New introductory Note. 128pp. 5 3/16 x 8 1/4. 0-486-27786-0

BASIC DRAWING, Louis Priscilla. Beginning with perspective, this commonsense manual progresses to the figure in movement, light and shade, anatomy, drapery, composition, trees and landscape, and outdoor sketching. Black-and-white illustrations throughout. 128pp. 8 3/8 x 11. 0-486-45815-6

THE BATTLES THAT CHANGED HISTORY, Fletcher Pratt. Historian profiles 16 crucial conflicts, ancient to modern, that changed the course of Western civilization. Gripping accounts of battles led by Alexander the Great, Joan of Arc, Ulysses S. Grant, other commanders. 27 maps. 352pp. 5⅜ x 8½. 0-486-41129-X

BEETHOVEN'S LETTERS, Ludwig van Beethoven. Edited by Dr. A. C. Kalischer. Features 457 letters to fellow musicians, friends, greats, patrons, and literary men. Reveals musical thoughts, quirks of personality, insights, and daily events. Includes 15 plates. 410pp. 5⅜ x 8½. 0-486-22769-3

BERNICE BOBS HER HAIR AND OTHER STORIES, F. Scott Fitzgerald. This brilliant anthology includes 6 of Fitzgerald's most popular stories: "The Diamond as Big as the Ritz," the title tale, "The Offshore Pirate," "The Ice Palace," "The Jelly Bean," and "May Day." 176pp. 5⅜ x 8½. 0-486-47049-0

BESLER'S BOOK OF FLOWERS AND PLANTS: 73 Full-Color Plates from Hortus Eystettensis, 1613, Basilius Besler. Here is a selection of magnificent plates from the *Hortus Eystettensis,* which vividly illustrated and identified the plants, flowers, and trees that thrived in the legendary German garden at Eichstätt. 80pp. 8⅜ x 11. 0-486-46005-3

THE BOOK OF KELLS, Edited by Blanche Cirker. Painstakingly reproduced from a rare facsimile edition, this volume contains full-page decorations, portraits, illustrations, plus a sampling of textual leaves with exquisite calligraphy and ornamentation. 32 full-color illustrations. 32pp. 9⅜ x 12¼. 0-486-24345-1

THE BOOK OF THE CROSSBOW: With an Additional Section on Catapults and Other Siege Engines, Ralph Payne-Gallwey. Fascinating study traces history and use of crossbow as military and sporting weapon, from Middle Ages to modern times. Also covers related weapons: balistas, catapults, Turkish bows, more. Over 240 illustrations. 400pp. 7¼ x 10⅛. 0-486-28720-3

THE BUNGALOW BOOK: Floor Plans and Photos of 112 Houses, 1910, Henry L. Wilson. Here are 112 of the most popular and economic blueprints of the early 20th century — plus an illustration or photograph of each completed house. A wonderful time capsule that still offers a wealth of valuable insights. 160pp. 8⅜ x 11. 0-486-45104-6

THE CALL OF THE WILD, Jack London. A classic novel of adventure, drawn from London's own experiences as a Klondike adventurer, relating the story of a heroic dog caught in the brutal life of the Alaska Gold Rush. Note. 64pp. 5 5/16 x 8¼. 0-486-26472-6

CANDIDE, Voltaire. Edited by Francois-Marie Arouet. One of the world's great satires since its first publication in 1759. Witty, caustic skewering of romance, science, philosophy, religion, government — nearly all human ideals and institutions. 112pp. 5 5/16 x 8¼. 0-486-26689-3

CELEBRATED IN THEIR TIME: Photographic Portraits from the George Grantham Bain Collection, Edited by Amy Pastan. With an Introduction by Michael Carlebach. Remarkable portrait gallery features 112 rare images of Albert Einstein, Charlie Chaplin, the Wright Brothers, Henry Ford, and other luminaries from the worlds of politics, art, entertainment, and industry. 128pp. 8⅜ x 11. 0-486-46754-6

CHARIOTS FOR APOLLO: The NASA History of Manned Lunar Spacecraft to 1969, Courtney G. Brooks, James M. Grimwood, and Loyd S. Swenson, Jr. This illustrated history by a trio of experts is the definitive reference on the Apollo spacecraft and lunar modules. It traces the vehicles' design, development, and operation in space. More than 100 photographs and illustrations. 576pp. 6¾ x 9¼. 0-486-46756-2

A CHRISTMAS CAROL, Charles Dickens. This engrossing tale relates Ebenezer Scrooge's ghostly journeys through Christmases past, present, and future and his ultimate transformation from a harsh and grasping old miser to a charitable and compassionate human being. 80pp. 5 3/16 x 8 1/4. 0-486-26865-9

COMMON SENSE, Thomas Paine. First published in January of 1776, this highly influential landmark document clearly and persuasively argued for American separation from Great Britain and paved the way for the Declaration of Independence. 64pp. 5 3/16 x 8 1/4. 0-486-29602-4

THE COMPLETE SHORT STORIES OF OSCAR WILDE, Oscar Wilde. Complete texts of "The Happy Prince and Other Tales," "A House of Pomegranates," "Lord Arthur Savile's Crime and Other Stories," "Poems in Prose," and "The Portrait of Mr. W. H." 208pp. 5 3/16 x 8 1/4. 0-486-45216-6

COMPLETE SONNETS, William Shakespeare. Over 150 exquisite poems deal with love, friendship, the tyranny of time, beauty's evanescence, death, and other themes in language of remarkable power, precision, and beauty. Glossary of archaic terms. 80pp. 5 3/16 x 8 1/4. 0-486-26686-9

THE COUNT OF MONTE CRISTO: Abridged Edition, Alexandre Dumas. Falsely accused of treason, Edmond Dantès is imprisoned in the bleak Chateau d'If. After a hair-raising escape, he launches an elaborate plot to extract a bitter revenge against those who betrayed him. 448pp. 5 3/16 x 8 1/4. 0-486-45643-9

CRAFTSMAN BUNGALOWS: Designs from the Pacific Northwest, Yoho & Merritt. This reprint of a rare catalog, showcasing the charming simplicity and cozy style of Craftsman bungalows, is filled with photos of completed homes, plus floor plans and estimated costs. An indispensable resource for architects, historians, and illustrators. 112pp. 10 x 7. 0-486-46875-5

CRAFTSMAN BUNGALOWS: 59 Homes from "The Craftsman," Edited by Gustav Stickley. Best and most attractive designs from Arts and Crafts Movement publication — 1903–1916 — includes sketches, photographs of homes, floor plans, descriptive text. 128pp. 8 1/4 x 11. 0-486-25829-7

CRIME AND PUNISHMENT, Fyodor Dostoyevsky. Translated by Constance Garnett. Supreme masterpiece tells the story of Raskolnikov, a student tormented by his own thoughts after he murders an old woman. Overwhelmed by guilt and terror, he confesses and goes to prison. 480pp. 5 3/16 x 8 1/4. 0-486-41587-2

THE DECLARATION OF INDEPENDENCE AND OTHER GREAT DOCUMENTS OF AMERICAN HISTORY: 1775-1865, Edited by John Grafton. Thirteen compelling and influential documents: Henry's "Give Me Liberty or Give Me Death," Declaration of Independence, The Constitution, Washington's First Inaugural Address, The Monroe Doctrine, The Emancipation Proclamation, Gettysburg Address, more. 64pp. 5 3/16 x 8 1/4. 0-486-41124-9

THE DESERT AND THE SOWN: Travels in Palestine and Syria, Gertrude Bell. "The female Lawrence of Arabia," Gertrude Bell wrote captivating, perceptive accounts of her travels in the Middle East. This intriguing narrative, accompanied by 160 photos, traces her 1905 sojourn in Lebanon, Syria, and Palestine. 368pp. 5 3/8 x 8 1/2. 0-486-46876-3

A DOLL'S HOUSE, Henrik Ibsen. Ibsen's best-known play displays his genius for realistic prose drama. An expression of women's rights, the play climaxes when the central character, Nora, rejects a smothering marriage and life in "a doll's house." 80pp. 5 3/16 x 8 1/4. 0-486-27062-9

DOOMED SHIPS: Great Ocean Liner Disasters, William H. Miller, Jr. Nearly 200 photographs, many from private collections, highlight tales of some of the vessels whose pleasure cruises ended in catastrophe: the *Morro Castle, Normandie, Andrea Doria, Europa,* and many others. 128pp. 8⅞ x 11¾. 0-486-45366-9

THE DORÉ BIBLE ILLUSTRATIONS, Gustave Doré. Detailed plates from the Bible: the Creation scenes, Adam and Eve, horrifying visions of the Flood, the battle sequences with their monumental crowds, depictions of the life of Jesus, 241 plates in all. 241pp. 9 x 12. 0-486-23004-X

DRAWING DRAPERY FROM HEAD TO TOE, Cliff Young. Expert guidance on how to draw shirts, pants, skirts, gloves, hats, and coats on the human figure, including folds in relation to the body, pull and crush, action folds, creases, more. Over 200 drawings. 48pp. 8¼ x 11. 0-486-45591-2

DUBLINERS, James Joyce. A fine and accessible introduction to the work of one of the 20th century's most influential writers, this collection features 15 tales, including a masterpiece of the short-story genre, "The Dead." 160pp. 5⁵⁄₁₆ x 8¼. 0-486-26870-5

EASY-TO-MAKE POP-UPS, Joan Irvine. Illustrated by Barbara Reid. Dozens of wonderful ideas for three-dimensional paper fun — from holiday greeting cards with moving parts to a pop-up menagerie. Easy-to-follow, illustrated instructions for more than 30 projects. 299 black-and-white illustrations. 96pp. 8⅜ x 11. 0-486-44622-0

EASY-TO-MAKE STORYBOOK DOLLS: A "Novel" Approach to Cloth Dollmaking, Sherralyn St. Clair. Favorite fictional characters come alive in this unique beginner's dollmaking guide. Includes patterns for Pollyanna, Dorothy from *The Wonderful Wizard of Oz,* Mary of *The Secret Garden,* plus easy-to-follow instructions, 263 black-and-white illustrations, and an 8-page color insert. 112pp. 8¼ x 11. 0-486-47360-0

EINSTEIN'S ESSAYS IN SCIENCE, Albert Einstein. Speeches and essays in accessible, everyday language profile influential physicists such as Niels Bohr and Isaac Newton. They also explore areas of physics to which the author made major contributions. 128pp. 5 x 8. 0-486-47011-3

EL DORADO: Further Adventures of the Scarlet Pimpernel, Baroness Orczy. A popular sequel to *The Scarlet Pimpernel,* this suspenseful story recounts the Pimpernel's attempts to rescue the Dauphin from imprisonment during the French Revolution. An irresistible blend of intrigue, period detail, and vibrant characterizations. 352pp. 5⁵⁄₁₆ x 8¼. 0-486-44026-5

ELEGANT SMALL HOMES OF THE TWENTIES: 99 Designs from a Competition, Chicago Tribune. Nearly 100 designs for five- and six-room houses feature New England and Southern colonials, Normandy cottages, stately Italianate dwellings, and other fascinating snapshots of American domestic architecture of the 1920s. 112pp. 9 x 12. 0-486-46910-7

THE ELEMENTS OF STYLE: The Original Edition, William Strunk, Jr. This is the book that generations of writers have relied upon for timeless advice on grammar, diction, syntax, and other essentials. In concise terms, it identifies the principal requirements of proper style and common errors. 64pp. 5⅜ x 8½. 0-486-44798-7

THE ELUSIVE PIMPERNEL, Baroness Orczy. Robespierre's revolutionaries find their wicked schemes thwarted by the heroic Pimpernel — Sir Percival Blakeney. In this thrilling sequel, Chauvelin devises a plot to eliminate the Pimpernel and his wife. 272pp. 5⁵⁄₁₆ x 8¼. 0-486-45464-9

AN ENCYCLOPEDIA OF BATTLES: Accounts of Over 1,560 Battles from 1479 B.C. to the Present, David Eggenberger. Essential details of every major battle in recorded history from the first battle of Megiddo in 1479 B.C. to Grenada in 1984. List of battle maps. 99 illustrations. 544pp. 6½ x 9¼. 0-486-24913-1

ENCYCLOPEDIA OF EMBROIDERY STITCHES, INCLUDING CREWEL, Marion Nichols. Precise explanations and instructions, clearly illustrated, on how to work chain, back, cross, knotted, woven stitches, and many more — 178 in all, including Cable Outline, Whipped Satin, and Eyelet Buttonhole. Over 1400 illustrations. 219pp. 8⅜ x 11¼. 0-486-22929-7

ENTER JEEVES: 15 Early Stories, P. G. Wodehouse. Splendid collection contains first 8 stories featuring Bertie Wooster, the deliciously dim aristocrat and Jeeves, his brainy, imperturbable manservant. Also, the complete Reggie Pepper (Bertie's prototype) series. 288pp. 5⅜ x 8½. 0-486-29717-9

ERIC SLOANE'S AMERICA: Paintings in Oil, Michael Wigley. With a Foreword by Mimi Sloane. Eric Sloane's evocative oils of America's landscape and material culture shimmer with immense historical and nostalgic appeal. This original hardcover collection gathers nearly a hundred of his finest paintings, with subjects ranging from New England to the American Southwest. 128pp. 10⅝ x 9. 0-486-46525-X

ETHAN FROME, Edith Wharton. Classic story of wasted lives, set against a bleak New England background. Superbly delineated characters in a hauntingly grim tale of thwarted love. Considered by many to be Wharton's masterpiece. 96pp. 5³⁄₁₆ x 8 ¼. 0-486-26690-7

THE EVERLASTING MAN, G. K. Chesterton. Chesterton's view of Christianity — as a blend of philosophy and mythology, satisfying intellect and spirit — applies to his brilliant book, which appeals to readers' heads as well as their hearts. 288pp. 5⅜ x 8½. 0-486-46036-3

THE FIELD AND FOREST HANDY BOOK, Daniel Beard. Written by a co-founder of the Boy Scouts, this appealing guide offers illustrated instructions for building kites, birdhouses, boats, igloos, and other fun projects, plus numerous helpful tips for campers. 448pp. 5³⁄₁₆ x 8¼. 0-486-46191-2

FINDING YOUR WAY WITHOUT MAP OR COMPASS, Harold Gatty. Useful, instructive manual shows would-be explorers, hikers, bikers, scouts, sailors, and survivalists how to find their way outdoors by observing animals, weather patterns, shifting sands, and other elements of nature. 288pp. 5⅜ x 8½. 0-486-40613-X

FIRST FRENCH READER: A Beginner's Dual-Language Book, Edited and Translated by Stanley Appelbaum. This anthology introduces 50 legendary writers — Voltaire, Balzac, Baudelaire, Proust, more — through passages from *The Red and the Black, Les Misérables, Madame Bovary,* and other classics. Original French text plus English translation on facing pages. 240pp. 5⅜ x 8½. 0-486-46178-5

FIRST GERMAN READER: A Beginner's Dual-Language Book, Edited by Harry Steinhauer. Specially chosen for their power to evoke German life and culture, these short, simple readings include poems, stories, essays, and anecdotes by Goethe, Hesse, Heine, Schiller, and others. 224pp. 5⅜ x 8½. 0-486-46179-3

FIRST SPANISH READER: A Beginner's Dual-Language Book, Angel Flores. Delightful stories, other material based on works of Don Juan Manuel, Luis Taboada, Ricardo Palma, other noted writers. Complete faithful English translations on facing pages. Exercises. 176pp. 5⅜ x 8½. 0-486-25810-6

FIVE ACRES AND INDEPENDENCE, Maurice G. Kains. Great back-to-the-land classic explains basics of self-sufficient farming. The one book to get. 95 illustrations. 397pp. 5⅜ x 8½. 0-486-20974-1

FLAGG'S SMALL HOUSES: Their Economic Design and Construction, 1922, Ernest Flagg. Although most famous for his skyscrapers, Flagg was also a proponent of the well-designed single-family dwelling. His classic treatise features innovations that save space, materials, and cost. 526 illustrations. 160pp. 9⅜ x 12¼. 0-486-45197-6

FLATLAND: A Romance of Many Dimensions, Edwin A. Abbott. Classic of science (and mathematical) fiction — charmingly illustrated by the author — describes the adventures of A. Square, a resident of Flatland, in Spaceland (three dimensions), Lineland (one dimension), and Pointland (no dimensions). 96pp. 5⁹⁄₁₆ x 8¼. 0-486-27263-X

FRANKENSTEIN, Mary Shelley. The story of Victor Frankenstein's monstrous creation and the havoc it caused has enthralled generations of readers and inspired countless writers of horror and suspense. With the author's own 1831 introduction. 176pp. 5⁹⁄₁₆ x 8¼. 0-486-28211-2

THE GARGOYLE BOOK: 572 Examples from Gothic Architecture, Lester Burbank Bridaham. Dispelling the conventional wisdom that French Gothic architectural flourishes were born of despair or gloom, Bridaham reveals the whimsical nature of these creations and the ingenious artisans who made them. 572 illustrations. 224pp. 8⅜ x 11. 0-486-44754-5

THE GIFT OF THE MAGI AND OTHER SHORT STORIES, O. Henry. Sixteen captivating stories by one of America's most popular storytellers. Included are such classics as "The Gift of the Magi," "The Last Leaf," and "The Ransom of Red Chief." Publisher's Note. 96pp. 5⁹⁄₁₆ x 8¼. 0-486-27061-0

THE GOETHE TREASURY: Selected Prose and Poetry, Johann Wolfgang von Goethe. Edited, Selected, and with an Introduction by Thomas Mann. In addition to his lyric poetry, Goethe wrote travel sketches, autobiographical studies, essays, letters, and proverbs in rhyme and prose. This collection presents outstanding examples from each genre. 368pp. 5⅜ x 8½. 0-486-44780-4

GREAT EXPECTATIONS, Charles Dickens. Orphaned Pip is apprenticed to the dirty work of the forge but dreams of becoming a gentleman — and one day finds himself in possession of "great expectations." Dickens' finest novel. 400pp. 5⁹⁄₁₆ x 8¼. 0-486-41586-4

GREAT WRITERS ON THE ART OF FICTION: From Mark Twain to Joyce Carol Oates, Edited by James Daley. An indispensable source of advice and inspiration, this anthology features essays by Henry James, Kate Chopin, Willa Cather, Sinclair Lewis, Jack London, Raymond Chandler, Raymond Carver, Eudora Welty, and Kurt Vonnegut, Jr. 192pp. 5⅜ x 8½. 0-486-45128-3

HAMLET, William Shakespeare. The quintessential Shakespearean tragedy, whose highly charged confrontations and anguished soliloquies probe depths of human feeling rarely sounded in any art. Reprinted from an authoritative British edition complete with illuminating footnotes. 128pp. 5⁹⁄₁₆ x 8¼. 0-486-27278-8

THE HAUNTED HOUSE, Charles Dickens. A Yuletide gathering in an eerie country retreat provides the backdrop for Dickens and his friends — including Elizabeth Gaskell and Wilkie Collins — who take turns spinning supernatural yarns. 144pp. 5⅜ x 8½. 0-486-46309-5

HEART OF DARKNESS, Joseph Conrad. Dark allegory of a journey up the Congo River and the narrator's encounter with the mysterious Mr. Kurtz. Masterly blend of adventure, character study, psychological penetration. For many, Conrad's finest, most enigmatic story. 80pp. 5 3/16 x 8 1/4. 0-486-26464-5

HENSON AT THE NORTH POLE, Matthew A. Henson. This thrilling memoir by the heroic African-American who was Peary's companion through two decades of Arctic exploration recounts a tale of danger, courage, and determination. "Fascinating and exciting." — *Commonweal.* 128pp. 5 3/8 x 8 1/2. 0-486-45472-X

HISTORIC COSTUMES AND HOW TO MAKE THEM, Mary Fernald and E. Shenton. Practical, informative guidebook shows how to create everything from short tunics worn by Saxon men in the fifth century to a lady's bustle dress of the late 1800s. 81 illustrations. 176pp. 5 3/8 x 8 1/2. 0-486-44906-8

THE HOUND OF THE BASKERVILLES, Arthur Conan Doyle. A deadly curse in the form of a legendary ferocious beast continues to claim its victims from the Baskerville family until Holmes and Watson intervene. Often called the best detective story ever written. 128pp. 5 3/16 x 8 1/4. 0-486-28214-7

THE HOUSE BEHIND THE CEDARS, Charles W. Chesnutt. Originally published in 1900, this groundbreaking novel by a distinguished African-American author recounts the drama of a brother and sister who "pass for white" during the dangerous days of Reconstruction. 208pp. 5 3/8 x 8 1/2. 0-486-46144-0

THE HUMAN FIGURE IN MOTION, Eadweard Muybridge. The 4,789 photographs in this definitive selection show the human figure — models almost all undraped — engaged in over 160 different types of action: running, climbing stairs, etc. 390pp. 7 7/8 x 10 5/8. 0-486-20204-6

THE IMPORTANCE OF BEING EARNEST, Oscar Wilde. Wilde's witty and buoyant comedy of manners, filled with some of literature's most famous epigrams, reprinted from an authoritative British edition. Considered Wilde's most perfect work. 64pp. 5 3/16 x 8 1/4. 0-486-26478-5

THE INFERNO, Dante Alighieri. Translated and with notes by Henry Wadsworth Longfellow. The first stop on Dante's famous journey from Hell to Purgatory to Paradise, this 14th-century allegorical poem blends vivid and shocking imagery with graceful lyricism. Translated by the beloved 19th-century poet, Henry Wadsworth Longfellow. 256pp. 5 3/16 x 8 1/4. 0-486-44288-8

JANE EYRE, Charlotte Brontë. Written in 1847, *Jane Eyre* tells the tale of an orphan girl's progress from the custody of cruel relatives to an oppressive boarding school and its culmination in a troubled career as a governess. 448pp. 5 3/16 x 8 1/4. 0-486-42449-9

JAPANESE WOODBLOCK FLOWER PRINTS, Tanigami Kônan. Extraordinary collection of Japanese woodblock prints by a well-known artist features 120 plates in brilliant color. Realistic images from a rare edition include daffodils, tulips, and other familiar and unusual flowers. 128pp. 11 x 8 1/4. 0-486-46442-3

JEWELRY MAKING AND DESIGN, Augustus F. Rose and Antonio Cirino. Professional secrets of jewelry making are revealed in a thorough, practical guide. Over 200 illustrations. 306pp. 5 3/8 x 8 1/2. 0-486-21750-7

JULIUS CAESAR, William Shakespeare. Great tragedy based on Plutarch's account of the lives of Brutus, Julius Caesar and Mark Antony. Evil plotting, ringing oratory, high tragedy with Shakespeare's incomparable insight, dramatic power. Explanatory footnotes. 96pp. 5 3/16 x 8 1/4. 0-486-26876-4

THE JUNGLE, Upton Sinclair. 1906 bestseller shockingly reveals intolerable labor practices and working conditions in the Chicago stockyards as it tells the grim story of a Slavic family that emigrates to America full of optimism but soon faces despair. 320pp. 5 3/16 x 8 1/4. 0-486-41923-1

THE KINGDOM OF GOD IS WITHIN YOU, Leo Tolstoy. The soul-searching book that inspired Gandhi to embrace the concept of passive resistance, Tolstoy's 1894 polemic clearly outlines a radical, well-reasoned revision of traditional Christian thinking. 352pp. 5 3/16 x 8 1/4. 0-486-45138-0

THE LADY OR THE TIGER?: and Other Logic Puzzles, Raymond M. Smullyan. Created by a renowned puzzle master, these whimsically themed challenges involve paradoxes about probability, time, and change; metapuzzles; and self-referentiality. Nineteen chapters advance in difficulty from relatively simple to highly complex. 1982 edition. 240pp. 5 3/8 x 8 1/2. 0-486-47027-X

LEAVES OF GRASS: The Original 1855 Edition, Walt Whitman. Whitman's immortal collection includes some of the greatest poems of modern times, including his masterpiece, "Song of Myself." Shattering standard conventions, it stands as an unabashed celebration of body and nature. 128pp. 5 3/16 x 8 1/4. 0-486-45676-5

LES MISÉRABLES, Victor Hugo. Translated by Charles E. Wilbour. Abridged by James K. Robinson. A convict's heroic struggle for justice and redemption plays out against a fiery backdrop of the Napoleonic wars. This edition features the excellent original translation and a sensitive abridgment. 304pp. 6 1/8 x 9 1/4. 0-486-45789-3

LILITH: A Romance, George MacDonald. In this novel by the father of fantasy literature, a man travels through time to meet Adam and Eve and to explore humanity's fall from grace and ultimate redemption. 240pp. 5 3/8 x 8 1/2. 0-486-46818-6

THE LOST LANGUAGE OF SYMBOLISM, Harold Bayley. This remarkable book reveals the hidden meaning behind familiar images and words, from the origins of Santa Claus to the fleur-de-lys, drawing from mythology, folklore, religious texts, and fairy tales. 1,418 illustrations. 784pp. 5 3/8 x 8 1/2. 0-486-44787-1

MACBETH, William Shakespeare. A Scottish nobleman murders the king in order to succeed to the throne. Tortured by his conscience and fearful of discovery, he becomes tangled in a web of treachery and deceit that ultimately spells his doom. 96pp. 5 3/16 x 8 1/4. 0-486-27802-6

MAKING AUTHENTIC CRAFTSMAN FURNITURE: Instructions and Plans for 62 Projects, Gustav Stickley. Make authentic reproductions of handsome, functional, durable furniture: tables, chairs, wall cabinets, desks, a hall tree, and more. Construction plans with drawings, schematics, dimensions, and lumber specs reprinted from 1900s *The Craftsman* magazine. 128pp. 8 1/8 x 11. 0-486-25000-8

MATHEMATICS FOR THE NONMATHEMATICIAN, Morris Kline. Erudite and entertaining overview follows development of mathematics from ancient Greeks to present. Topics include logic and mathematics, the fundamental concept, differential calculus, probability theory, much more. Exercises and problems. 641pp. 5 3/8 x 8 1/2. 0-486-24823-2

MEMOIRS OF AN ARABIAN PRINCESS FROM ZANZIBAR, Emily Ruete. This 19th-century autobiography offers a rare inside look at the society surrounding a sultan's palace. A real-life princess in exile recalls her vanished world of harems, slave trading, and court intrigues. 288pp. 5 3/8 x 8 1/2. 0-486-47121-7

THE METAMORPHOSIS AND OTHER STORIES, Franz Kafka. Excellent new English translations of title story (considered by many critics Kafka's most perfect work), plus "The Judgment," "In the Penal Colony," "A Country Doctor," and "A Report to an Academy." Note. 96pp. 5 3/16 x 8 1/4. 0-486-29030-1

MICROSCOPIC ART FORMS FROM THE PLANT WORLD, R. Anheisser. From undulating curves to complex geometrics, a world of fascinating images abound in this classic, illustrated survey of microscopic plants. Features 400 detailed illustrations of nature's minute but magnificent handiwork. The accompanying CD-ROM includes all of the images in the book. 128pp. 9 x 9. 0-486-46013-4

A MIDSUMMER NIGHT'S DREAM, William Shakespeare. Among the most popular of Shakespeare's comedies, this enchanting play humorously celebrates the vagaries of love as it focuses upon the intertwined romances of several pairs of lovers. Explanatory footnotes. 80pp. 5 3/16 x 8 1/4. 0-486-27067-X

THE MONEY CHANGERS, Upton Sinclair. Originally published in 1908, this cautionary novel from the author of *The Jungle* explores corruption within the American system as a group of power brokers joins forces for personal gain, triggering a crash on Wall Street. 192pp. 5 3/8 x 8 1/2. 0-486-46917-4

THE MOST POPULAR HOMES OF THE TWENTIES, William A. Radford. With a New Introduction by Daniel D. Reiff. Based on a rare 1925 catalog, this architectural showcase features floor plans, construction details, and photos of 26 homes, plus articles on entrances, porches, garages, and more. 250 illustrations, 21 color plates. 176pp. 8 3/8 x 11. 0-486-47028-8

MY 66 YEARS IN THE BIG LEAGUES, Connie Mack. With a New Introduction by Rich Westcott. A Founding Father of modern baseball, Mack holds the record for most wins — and losses — by a major league manager. Enhanced by 70 photographs, his warmhearted autobiography is populated by many legends of the game. 288pp. 5 3/8 x 8 1/2. 0-486-47184-5

NARRATIVE OF THE LIFE OF FREDERICK DOUGLASS, Frederick Douglass. Douglass's graphic depictions of slavery, harrowing escape to freedom, and life as a newspaper editor, eloquent orator, and impassioned abolitionist. 96pp. 5 3/16 x 8 1/4. 0-486-28499-9

THE NIGHTLESS CITY: Geisha and Courtesan Life in Old Tokyo, J. E. de Becker. This unsurpassed study from 100 years ago ventured into Tokyo's red-light district to survey geisha and courtesan life and offer meticulous descriptions of training, dress, social hierarchy, and erotic practices. 49 black-and-white illustrations; 2 maps. 496pp. 5 3/8 x 8 1/2. 0-486-45563-7

THE ODYSSEY, Homer. Excellent prose translation of ancient epic recounts adventures of the homeward-bound Odysseus. Fantastic cast of gods, giants, cannibals, sirens, other supernatural creatures — true classic of Western literature. 256pp. 5 3/16 x 8 1/4. 0-486-40654-7

OEDIPUS REX, Sophocles. Landmark of Western drama concerns the catastrophe that ensues when King Oedipus discovers he has inadvertently killed his father and married his mother. Masterly construction, dramatic irony. Explanatory footnotes. 64pp. 5 3/16 x 8 1/4. 0-486-26877-2

ONCE UPON A TIME: The Way America Was, Eric Sloane. Nostalgic text and drawings brim with gentle philosophies and descriptions of how we used to live — self-sufficiently — on the land, in homes, and among the things built by hand. 44 line illustrations. 64pp. 8 3/8 x 11. 0-486-44411-2

ONE OF OURS, Willa Cather. The Pulitzer Prize–winning novel about a young Nebraskan looking for something to believe in. Alienated from his parents, rejected by his wife, he finds his destiny on the bloody battlefields of World War I. 352pp. 5⅜ x 8¼. 0-486-45599-8

ORIGAMI YOU CAN USE: 27 Practical Projects, Rick Beech. Origami models can be more than decorative, and this unique volume shows how! The 27 practical projects include a CD case, frame, napkin ring, and dish. Easy instructions feature 400 two-color illustrations. 96pp. 8¼ x 11. 0-486-47057-1

OTHELLO, William Shakespeare. Towering tragedy tells the story of a Moorish general who earns the enmity of his ensign Iago when he passes him over for a promotion. Masterly portrait of an archvillain. Explanatory footnotes. 112pp. 5⅜ x 8¼. 0-486-29097-2

PARADISE LOST, John Milton. Notes by John A. Himes. First published in 1667, *Paradise Lost* ranks among the greatest of English literature's epic poems. It's a sublime retelling of Adam and Eve's fall from grace and expulsion from Eden. Notes by John A. Himes. 480pp. 5⅜ x 8¼. 0-486-44287-X

PASSING, Nella Larsen. Married to a successful physician and prominently ensconced in society, Irene Redfield leads a charmed existence — until a chance encounter with a childhood friend who has been "passing for white." 112pp. 5⅜ x 8½. 0-486-43713-2

PERSPECTIVE DRAWING FOR BEGINNERS, Len A. Doust. Doust carefully explains the roles of lines, boxes, and circles, and shows how visualizing shapes and forms can be used in accurate depictions of perspective. One of the most concise introductions available. 33 illustrations. 64pp. 5⅜ x 8½. 0-486-45149-6

PERSPECTIVE MADE EASY, Ernest R. Norling. Perspective is easy; yet, surprisingly few artists know the simple rules that make it so. Remedy that situation with this simple, step-by-step book, the first devoted entirely to the topic. 256 illustrations. 224pp. 5⅜ x 8½. 0-486-40473-0

THE PICTURE OF DORIAN GRAY, Oscar Wilde. Celebrated novel involves a handsome young Londoner who sinks into a life of depravity. His body retains perfect youth and vigor while his recent portrait reflects the ravages of his crime and sensuality. 176pp. 5⅜ x 8¼. 0-486-27807-7

PRIDE AND PREJUDICE, Jane Austen. One of the most universally loved and admired English novels, an effervescent tale of rural romance transformed by Jane Austen's art into a witty, shrewdly observed satire of English country life. 272pp. 5⅜ x 8¼. 0-486-28473-5

THE PRINCE, Niccolò Machiavelli. Classic, Renaissance-era guide to acquiring and maintaining political power. Today, nearly 500 years after it was written, this calculating prescription for autocratic rule continues to be much read and studied. 80pp. 5⅜ x 8¼. 0-486-27274-5

QUICK SKETCHING, Carl Cheek. A perfect introduction to the technique of "quick sketching." Drawing upon an artist's immediate emotional responses, this is an extremely effective means of capturing the essential form and features of a subject. More than 100 black-and-white illustrations throughout. 48pp. 11 x 8¼. 0-486-46608-6

RANCH LIFE AND THE HUNTING TRAIL, Theodore Roosevelt. Illustrated by Frederic Remington. Beautifully illustrated by Remington, Roosevelt's celebration of the Old West recounts his adventures in the Dakota Badlands of the 1880s, from roundups to Indian encounters to hunting bighorn sheep. 208pp. 6¼ x 9¼. 0-486-47340-6

Browse over 9,000 books at www.doverpublications.com

THE RED BADGE OF COURAGE, Stephen Crane. Amid the nightmarish chaos of a Civil War battle, a young soldier discovers courage, humility, and, perhaps, wisdom. Uncanny re-creation of actual combat. Enduring landmark of American fiction. 112pp. 5 3/16 x 8 1/4. 0-486-26465-3

RELATIVITY SIMPLY EXPLAINED, Martin Gardner. One of the subject's clearest, most entertaining introductions offers lucid explanations of special and general theories of relativity, gravity, and spacetime, models of the universe, and more. 100 illustrations. 224pp. 5 3/8 x 8 1/2. 0-486-29315-7

REMBRANDT DRAWINGS: 116 Masterpieces in Original Color, Rembrandt van Rijn. This deluxe hardcover edition features drawings from throughout the Dutch master's prolific career. Informative captions accompany these beautifully reproduced landscapes, biblical vignettes, figure studies, animal sketches, and portraits. 128pp. 8 3/8 x 11. 0-486-46149-1

THE ROAD NOT TAKEN AND OTHER POEMS, Robert Frost. A treasury of Frost's most expressive verse. In addition to the title poem: "An Old Man's Winter Night," "In the Home Stretch," "Meeting and Passing," "Putting in the Seed," many more. All complete and unabridged. 64pp. 5 3/16 x 8 1/4. 0-486-27550-7

ROMEO AND JULIET, William Shakespeare. Tragic tale of star-crossed lovers, feuding families and timeless passion contains some of Shakespeare's most beautiful and lyrical love poetry. Complete, unabridged text with explanatory footnotes. 96pp. 5 3/16 x 8 1/4. 0-486-27557-4

SANDITON AND THE WATSONS: Austen's Unfinished Novels, Jane Austen. Two tantalizing incomplete stories revisit Austen's customary milieu of courtship and venture into new territory, amid guests at a seaside resort. Both are worth reading for pleasure and study. 112pp. 5 3/8 x 8 1/2. 0-486-45793-1

THE SCARLET LETTER, Nathaniel Hawthorne. With stark power and emotional depth, Hawthorne's masterpiece explores sin, guilt, and redemption in a story of adultery in the early days of the Massachusetts Colony. 192pp. 5 3/16 x 8 1/4. 0-486-28048-9

THE SEASONS OF AMERICA PAST, Eric Sloane. Seventy-five illustrations depict cider mills and presses, sleds, pumps, stump-pulling equipment, plows, and other elements of America's rural heritage. A section of old recipes and household hints adds additional color. 160pp. 8 3/8 x 11. 0-486-44220-9

SELECTED CANTERBURY TALES, Geoffrey Chaucer. Delightful collection includes the General Prologue plus three of the most popular tales: "The Knight's Tale," "The Miller's Prologue and Tale," and "The Wife of Bath's Prologue and Tale." In modern English. 144pp. 5 3/16 x 8 1/4. 0-486-28241-4

SELECTED POEMS, Emily Dickinson. Over 100 best-known, best-loved poems by one of America's foremost poets, reprinted from authoritative early editions. No comparable edition at this price. Index of first lines. 64pp. 5 3/16 x 8 1/4. 0-486-26466-1

SIDDHARTHA, Hermann Hesse. Classic novel that has inspired generations of seekers. Blending Eastern mysticism and psychoanalysis, Hesse presents a strikingly original view of man and culture and the arduous process of self-discovery, reconciliation, harmony, and peace. 112pp. 5 3/16 x 8 1/4. 0-486-40653-9

SKETCHING OUTDOORS, Leonard Richmond. This guide offers beginners step-by-step demonstrations of how to depict clouds, trees, buildings, and other outdoor sights. Explanations of a variety of techniques include shading and constructional drawing. 48pp. 11 x 8 1/4. 0-486-46922-0

SMALL HOUSES OF THE FORTIES: With Illustrations and Floor Plans, Harold E. Group. 56 floor plans and elevations of houses that originally cost less than $15,000 to build. Recommended by financial institutions of the era, they range from Colonials to Cape Cods. 144pp. 8⅜ x 11. 0-486-45598-X

SOME CHINESE GHOSTS, Lafcadio Hearn. Rooted in ancient Chinese legends, these richly atmospheric supernatural tales are recounted by an expert in Oriental lore. Their originality, power, and literary charm will captivate readers of all ages. 96pp. 5⅜ x 8½. 0-486-46306-0

SONGS FOR THE OPEN ROAD: Poems of Travel and Adventure, Edited by The American Poetry & Literacy Project. More than 80 poems by 50 American and British masters celebrate real and metaphorical journeys. Poems by Whitman, Byron, Millay, Sandburg, Langston Hughes, Emily Dickinson, Robert Frost, Shelley, Tennyson, Yeats, many others. Note. 80pp. 5⁹⁄₁₆ x 8¼. 0-486-40646-6

SPOON RIVER ANTHOLOGY, Edgar Lee Masters. An American poetry classic, in which former citizens of a mythical midwestern town speak touchingly from the grave of the thwarted hopes and dreams of their lives. 144pp. 5⁹⁄₁₆ x 8¼. 0-486-27275-3

STAR LORE: Myths, Legends, and Facts, William Tyler Olcott. Captivating retellings of the origins and histories of ancient star groups include Pegasus, Ursa Major, Pleiades, signs of the zodiac, and other constellations. "Classic." — *Sky & Telescope.* 58 illustrations. 544pp. 5⅜ x 8½. 0-486-43581-4

THE STRANGE CASE OF DR. JEKYLL AND MR. HYDE, Robert Louis Stevenson. This intriguing novel, both fantasy thriller and moral allegory, depicts the struggle of two opposing personalities — one essentially good, the other evil — for the soul of one man. 64pp. 5⁹⁄₁₆ x 8¼. 0-486-26688-5

SURVIVAL HANDBOOK: The Official U.S. Army Guide, Department of the Army. This special edition of the Army field manual is geared toward civilians. An essential companion for campers and all lovers of the outdoors, it constitutes the most authoritative wilderness guide. 288pp. 5⁹⁄₁₆ x 8¼. 0-486-46184-X

A TALE OF TWO CITIES, Charles Dickens. Against the backdrop of the French Revolution, Dickens unfolds his masterpiece of drama, adventure, and romance about a man falsely accused of treason. Excitement and derring-do in the shadow of the guillotine. 304pp. 5⁹⁄₁₆ x 8¼. 0-486-40651-2

TEN PLAYS, Anton Chekhov. *The Sea Gull, Uncle Vanya, The Three Sisters, The Cherry Orchard,* and *Ivanov,* plus 5 one-act comedies: *The Anniversary, An Unwilling Martyr, The Wedding, The Bear,* and *The Proposal.* 336pp. 5⁹⁄₁₆ x 8¼. 0-486-46560-8

THE FLYING INN, G. K. Chesterton. Hilarious romp in which pub owner Humphrey Hump and friend take to the road in a donkey cart filled with rum and cheese, inveighing against Prohibition and other "oppressive forms of modernity." 320pp. 5⅜ x 8½. 0-486-41910-X

THIRTY YEARS THAT SHOOK PHYSICS: The Story of Quantum Theory, George Gamow. Lucid, accessible introduction to the influential theory of energy and matter features careful explanations of Dirac's anti-particles, Bohr's model of the atom, and much more. Numerous drawings. 1966 edition. 240pp. 5⅜ x 8½. 0-486-24895-X

TREASURE ISLAND, Robert Louis Stevenson. Classic adventure story of a perilous sea journey, a mutiny led by the infamous Long John Silver, and a lethal scramble for buried treasure — seen through the eyes of cabin boy Jim Hawkins. 160pp. 5⁹⁄₁₆ x 8¼. 0-486-27559-0

THE TRIAL, Franz Kafka. Translated by David Wyllie. From its gripping first sentence onward, this novel exemplifies the term "Kafkaesque." Its darkly humorous narrative recounts a bank clerk's entrapment in a bureaucratic maze, based on an undisclosed charge. 176pp. 5 3/16 x 8 1/4. 0-486-47061-X

THE TURN OF THE SCREW, Henry James. Gripping ghost story by great novelist depicts the sinister transformation of 2 innocent children into flagrant liars and hypocrites. An elegantly told tale of unspoken horror and psychological terror. 96pp. 5 3/16 x 8 1/4. 0-486-26684-2

UP FROM SLAVERY, Booker T. Washington. Washington (1856-1915) rose to become the most influential spokesman for African-Americans of his day. In this eloquently written book, he describes events in a remarkable life that began in bondage and culminated in worldwide recognition. 160pp. 5 3/16 x 8 1/4. 0-486-28738-6

VICTORIAN HOUSE DESIGNS IN AUTHENTIC FULL COLOR: 75 Plates from the "Scientific American – Architects and Builders Edition," 1885-1894, Edited by Blanche Cirker. Exquisitely detailed, exceptionally handsome designs for an enormous variety of attractive city dwellings, spacious suburban and country homes, charming "cottages" and other structures — all accompanied by perspective views and floor plans. 80pp. 9 1/4 x 12 1/4. 0-486-29438-2

VILLETTE, Charlotte Brontë. Acclaimed by Virginia Woolf as "Brontë's finest novel," this moving psychological study features a remarkably modern heroine who abandons her native England for a new life as a schoolteacher in Belgium. 480pp. 5 3/16 x 8 1/4. 0-486-45557-2

THE VOYAGE OUT, Virginia Woolf. A moving depiction of the thrills and confusion of youth, Woolf's acclaimed first novel traces a shipboard journey to South America for a captivating exploration of a woman's growing self-awareness. 288pp. 5 3/16 x 8 1/4. 0-486-45005-8

WALDEN; OR, LIFE IN THE WOODS, Henry David Thoreau. Accounts of Thoreau's daily life on the shores of Walden Pond outside Concord, Massachusetts, are interwoven with musings on the virtues of self-reliance and individual freedom, on society, government, and other topics. 224pp. 5 3/16 x 8 1/4. 0-486-28495-6

WILD PILGRIMAGE: A Novel in Woodcuts, Lynd Ward. Through startling engravings shaded in black and red, Ward wordlessly tells the story of a man trapped in an industrial world, struggling between the grim reality around him and the fantasies his imagination creates. 112pp. 6 1/8 x 9 1/4. 0-486-46583-7

WILLY POGÁNY REDISCOVERED, Willy Pogány. Selected and Edited by Jeff A. Menges. More than 100 color and black-and-white Art Nouveau–style illustrations from fairy tales and adventure stories include scenes from Wagner's "Ring" cycle, *The Rime of the Ancient Mariner, Gulliver's Travels,* and *Faust.* 144pp. 8 3/8 x 11. 0-486-47046-6

WOOLLY THOUGHTS: Unlock Your Creative Genius with Modular Knitting, Pat Ashforth and Steve Plummer. Here's the revolutionary way to knit — easy, fun, and foolproof! Beginners and experienced knitters need only master a single stitch to create their own designs with patchwork squares. More than 100 illustrations. 128pp. 6 1/2 x 9 1/4. 0-486-46084-3

WUTHERING HEIGHTS, Emily Brontë. Somber tale of consuming passions and vengeance — played out amid the lonely English moors — recounts the turbulent and tempestuous love story of Cathy and Heathcliff. Poignant and compelling. 256pp. 5 3/16 x 8 1/4. 0-486-29256-8

Browse over 9,000 books at www.doverpublications.com